Generis

PUBLISHING

A Half-Century Research Footpath in Statistical Physics

Alberto Robledo

Carlos Velarde

Title: **A Half-Century Research Footpath in Statistical Physics**

ISBN: 979-8-89248-943-0

Author: Alberto Robledo, Carlos Velarde

Cover image: www.pixabay.com

Publisher: Generis Publishing
Online orders: www.generis-publishing.com
Contact email: info@generis-publishing.com

Contents

Preface

We give an abridged account of a continued string of studies in condensed matter physics and in complex systems that span five decades. We provide links to access abstracts and full texts of a selected list of publications. The studies were carried out within a framework of methods and models, some developed in situ, of stochastic processes, statistical mechanics and nonlinear dynamics. The topics, techniques and outcomes reflect evolving interests of the community but also show a particular character that privileges the use of analogies or unusual viewpoints that unite the studies in distinctive ways. We place the studies into four collections according to the main underlying approach: stochastic processes, density functional theory, nonlinear dynamics, and generalized statistical mechanics. In the first group we include the following: i) Random walks for fluid correlations. ii) Random walks for electronic band structures. iii) Trapping of multiple walkers. iv) Renormalization group and entropy for Weierstrass walks. v) Statistical-mechanical analogy for renewal processes. vi) Phase transitions along time in correlated renewal processes. In the second group: i) Density functionals and Widom's particle insertion method. ii) Liquid to solid transitions in hard-core model systems. iii) Nucleation and spinodal decomposition. iv) Global wetting phase diagrams. v) Lattice models for complex fluids. vi) Anomalous micellar solution criticality. vii) Complex fluids under confinement. viii) Density functionals for curved interfaces. ix) Curvature interfacial transitions. x) Line tension and wetting. xi) Classical and quantum mechanical analogs built in density functional theory. xii) Phase behavior and pairing mechanism for two-dimensional superconductors. In the third group: i) Critical fluctuations and the route out of chaos. ii) Glassy dynamics at the noise-perturbed onset of chaos. iii) Localization transition as a bifurcation. iv) Pesin identity at the Feigenbaum attractor. v) Renormalization group for chaotic attractors. vi) Self-organization along the period-doubling cascade. vii) Universality classes of rank distributions. viii) Complex network view of the routes to chaos. ix) Chaos in discrete time game theory. x) Equivalence of paradigms (edge of chaos and criticality). xi) Allometry in biology and human activities. xii) Nested systems. In the fourth group: i) A new statistical mechanics. ii) How, why and when q-statistics is pertinent. In this last group we refer to our most recent contribution that underlies all the topics of the third group: The answer to the long-standing question about the limit of validity of ordinary statistical mechanics and the suitability of Tsallis statistics. We discuss the body of knowledge created by these research lines, indicate unsuspected connections beetween these investigations and point out perspectives for future developments.

Introduction

Half a century ago the frontiers of statistical physics were about to expand considerably and conquer new territories. Amongst them, and most visibly, the understanding of the scaling laws of critical phenomena exposed a few years earlier [1]. The advancement was made possible via the development of the novel research tool known as Renormalization Group [2]. Concurrently, and with features parallel to those of criticality, the nonlinear dynamical discovery of the routes to deterministic chaos from regular to irregular behavior took place [3]. Alongside, powerful methods for stochastic processes were just beginning to be applied to random walks and renewal processes associated with long-tailed distributions as opposed to the customary normal case [4]. Equally, at that time new statistical-mechanical techniques made possible a shift of focus from the study of simple homogeneous states towards more complicated situations. Contemporarily, studies of phase transitions in various different settings such as those occurring within surfaces became feasible, and the field of complex fluids, also referred to as soft condensed matter, grew to be differentiated from statistical mechanics of polymers [5]. We shall not describe here, in general, the evolution of statistical physics in the subsequent decades, but information on this can be obtained from different sources, such as, for example, the proceedings of the International Conferences on Statistical Physics (Statphys) that have occurred mostly with three-year periodicity over this time [6].

Five decades ago the Mexican research community in statistical physics could be practically counted with the fingers of one hand, but it was already undergoing a strong process of growth, with new participants naturally curious and positively influenced by the larger international community interests. By the turn of the century, half way through the 50-year period involved in this account, the Mexican community in the field had increased and matured very considerably. This was reflected by the organization and hosting of the Statphys Conference in Cancun, Mexico, in 2001 [7], the first to be held in a Spanish speaking country of these long series of flagship conferences. In the present day, the Mexican community in the field has developed much further and has become a large and vigorous research group spread into an array of topics that represent the current research pursuits in statistical physics and in frontier interdisciplinary science.

Here we give an account of an uninterrupted trajectory of studies based in Mexico and which covers this long period of development. The studies started by linking stochastic processes, such as random walks, to fluid or electronic structures, then the trapping of multiple walkers was considered, and much later, these studies led to a

Renormalization Group view for long-tailed distributions, and to renewal processes undergoing phase change along time. Another, larger set of studies, mainly based on the free energy density functional approach, produced descriptions for liquid-solid transitions, interfacial transitions, and kinetics of phase change. Subsequently, models were constructed for micellar solutions, microemulsions and magnetic alloys, the properties of various types of confined systems, such as nematics and mixtures of quiral molecules, were determined, and the understanding of magnetism and superconductivity phase behavior in quasi-two-dimensional systems was advanced. Latterly, through an additional set of studies, important features were uncovered on the dynamics at the onset of chaos in nonlinear systems, which unexpectedly connected to difficult problems in condensed matter physics, such as the dynamics close to glass formation, localization phenomena and critical fluctuations. But also, the same set of dynamical properties has been used for the design of models for complex systems phenomena including self-organization. Finally, the accumulated experience gained through the studies involving density functional theory and iterated-time nonlinear low-dimensional dynamics led to a promising advance regarding an enduring question about a generalization of statistical mechanics, the so-called q-statistics.

The studies are allocated into thirtytwo groups and these in turn into four main parts. Each part has a preamble followed by its sections (the groups of studies in the part). Each section is given an appropriate informative title and it is followed by a numbered collection of links to abstracts and references to selected advances that represent the focused subject. A particular selected advance is located by the part (I, II, III or IV), its section (i, ii, iii, ...), and its explanatory comment and links (1,2,3,...). We end our account with a summary of the most relevant developments and an outlook of possible future work.

A critical assessment is provided for each of the 32 topics. The assessments constitute the main informative content of this review, they indicate the basic advances accomplished but also their limitations; the novelty or uniqueness of the studies, then too the use of preceding knowledge and, or, previously developed methodologies and models. The articles referenced in each assessment contain more technical detail about these developments and provide specific information already in the literature. The statistical physics developments considered are in this case not all recent but extended throughout several decades, not global but restricted to a specific country, and not covering all topics but limited to a particular experience.

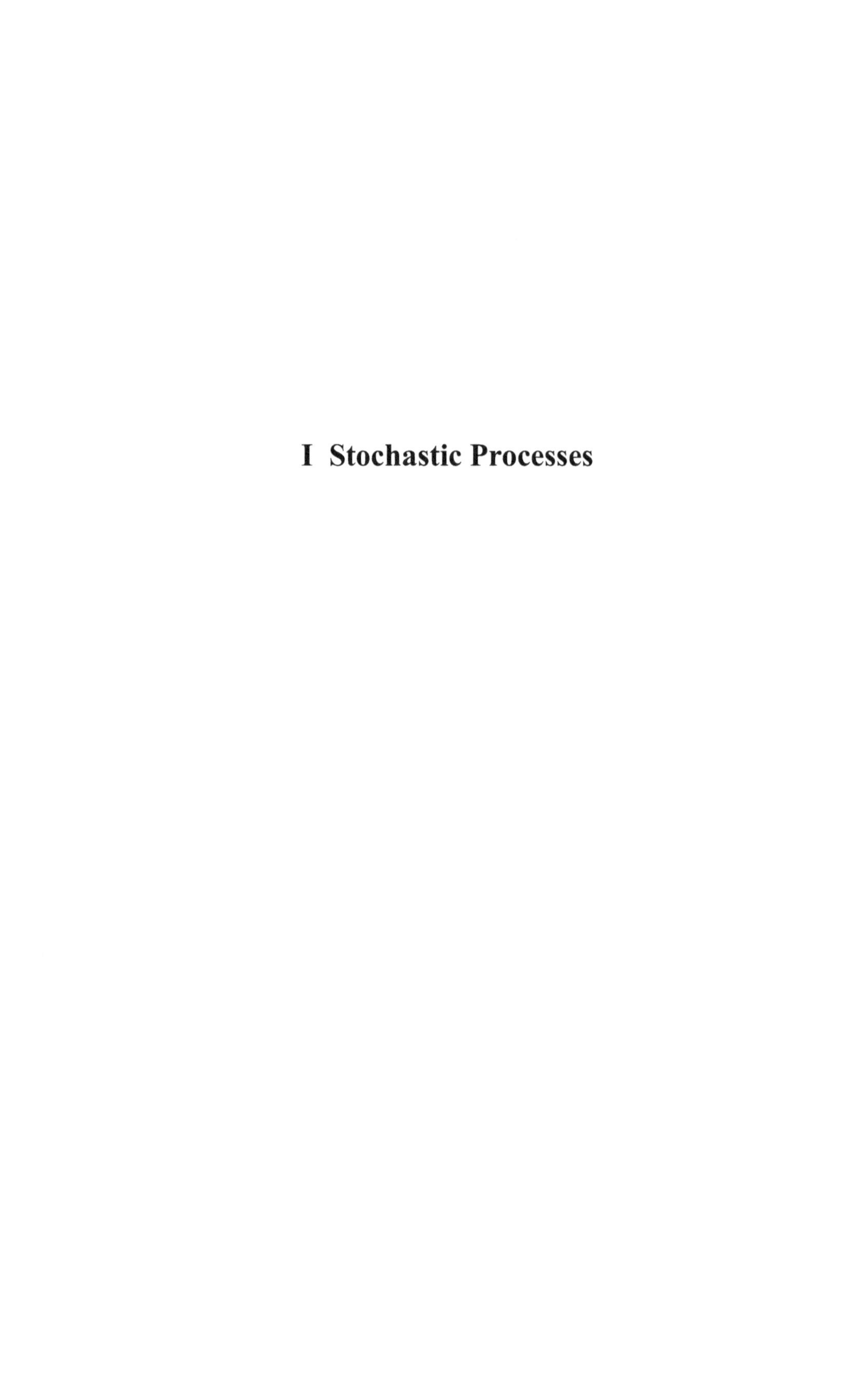

I Stochastic Processes

Random walk analogs. Elliot Montroll advanced a particularly efficient and elegant format to study random walks and renewal processes [8, 9]. The linearity property of independent identically distributed (iid) events (on regular lattices when spatial positions are required) allowed for a prominent role for the generating functions of the basic sets of probabilities. Once determined, Fourier, Laplace or z-transforms, their inverses, and use of differentiation when necessary, facilitates access to all conceivable properties [8, 9]. Recurrence relations are expressible as convolutions. First passage times become trapping events and traps can be arranged to form lines, surfaces, etc. Distinction between time and number of events along a sequence of steps leads to a fitting continuous time random walk description.

How to make use of these stochastic processes in other fields? First, we established analogies via the identification of convolutions present in the basic expressions. One case is the Ornstein-Zernicke relation, a common starting point for the development of integral equation approaches for correlation functions descriptive of equilibrium fluid structures. Another one is the Green function relation for tight-binding electronic band structures. Secondly, there is the requirement for primary events in photosynthesis to be triggered, only after the trapping at a reaction site of more than one energy packet. This led to the study of first passage times for multiple walkers.

The elegant Weierstrass walk, designed by Montroll and colleagues [10] to observe the emergence in the appropriate limits of either the Gaussian or the Levy distributions, has a rich set of scaling properties reminiscent of those of the nontrivial fixed points of the Renormalization Group (RG) when applied to critical phenomena in thermal systems. It was our interest to generalize this specific walk into an infinite family of them, that took the form of a universality class, to see if it revealed more features of this likeness, such as the flow to trivial fixed points. Moreover, it presented the opportunity of calculating entropies for the entire family of step distributions, and observe if the associated fixed points represent entropy extrema. In this way it was possible to reveal a hidden connection between the RG technique and entropy optimization [4].

In the case of renewal processes it caught our attention the fact that the expression for a key quantity resembled that of a grand canonical partition function in equilibrium statistical mechanics. This expression is that for the Laplace transform of the generating function for the probability density for the occurrence of n events at time t. This identification would imply that the probability density itself and its Laplace transform would take the place, respectively, of micro-canonical and canonical partition functions. The importance of the corroboration of this strict analogy is that there is a one to one correspondence between each specific renewal processes and a thermal system model with two fields, like a magnet or one-component fluid. The familiar case of independent identically distributed events is equivalent to the ideal gas but correlated event processes associate with more interesting statistical-mechanical systems. Amongst them we chose to look at the occurrence of phase transitions along time evolution. Recently, based on these developments for renewal processes, we constructed a model for cascades of events with application to pandemic spreading under population confinement.

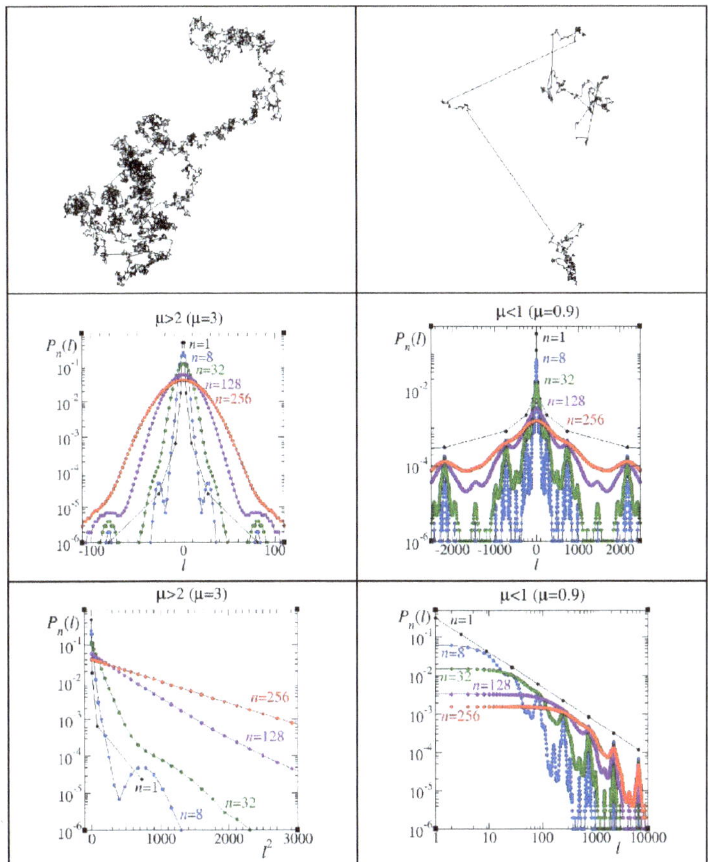

Figure 1: Random walks. The top panels illustrate trajectories for random walks with steps of varying lengths for which the balance from short to long step lengths is governed by a parameter μ. The middle panels show the probability distribution $P_n(l)$ for a walker to be at a position l after n steps for different values of n. Two different values of μ were chosen above and below a critical threshold $\mu_c = 2$. The bottom panels show the same probabilities in the middle panels, in one case in logarithmic scale for P and quadratic scale for l, and in the other case in logarithmic scales for both P and l, to corroborate the evolution towards Gaussian and Levy stationary distributions in the limit $n \to \infty$ for the chosen values of μ. Studies related to these figure panels are found in Section I.iv.

I.i Random walks and fluid correlations

The ingredients for this set of studies originate from two different sources. One of them the random walk generating functions for the probabilities of a walker position after a given number of steps. For symmetric nearest-neighbor step walks on infinite regular lattices these expressions were known [9] for several lattice dimensions and geometries. The other constituent was the Ornstein-Zernike equation relating pair correlation functions, an equation that to be useful is necessary to be complemented by approximate closure relations such as the Percus-Yevick approximation [11]. Both expressions involve convolutions. A general analogy between the two was developed [I.i.1, I.i.2] and was subsequently used to study hard-core lattice gases and the lattice version of the van der Waals model. The outcome was a description of order-disorder transitions [I.i.3], decay of fluid correlations [I.i.4], an extension to nonequilibrium of the Ornstein-Zernike relation [I.i.5], and a random walk version of phase coexistence [I.i.6]. Independence of events, linearity and lattice regularity were (and are) the limitations of this approach. There is no parallel version or development of this unusual analogy between stochastic processes and equilibrium statistical mechanics, and has remained basically unnoticed except for sporadic readers in recent years.

I.i.1 Random-walk theory and Ornstein-Zernike systems with extended-core potentials. A. Robledo, I.E. Farquhar. *J. Chem. Phys.* **61**, 1594 (1974) https://doi.org/10.1063/1.1682143

I.i.2 Random-walk theory and correlation functions in classical statistical mechanics. A. Robledo, I.E. Farquhar. *Physica A* **84**, 435 (1976) https://doi.org/10.1016/0378-4371(76)90097-2

I.i.3 Random-walk theory, ordered phases in lattice gas systems. A. Robledo, I.E. Farquhar. *Physica A* **84**, 449 (1976) https://doi.org/10.1016/0378-4371(76)90098-4

I.i.4 Random-walk theory and the decay of pair correlations in Ornstein-Zernike lattice systems. A. Robledo, I.E. Farquhar. *Physica A* **84**, 472 (1976) https://doi.org/10.1016/0378-4371(76)90099-6

I.i.5 On the relationship between continuous time random walks and the nonequilibrium Ornstein-Zernike equation. A.B. Budgor, A. Robledo. *Physica A* **85**, 329 (1976) https://doi.org/10.1016/0378-4371(76)90053-4

I.i.6 Random walks, the Ornstein-Zernike equation and the condensation of a 1-dimensional lattice gas. A. Robledo, A.B. Budgor. *Am. J. Phys.* **46**, 998 (1978) https://doi.org/10.1119/1.11491

I.ii Random walks and electronic band structures

As in the previous case two different sources were used to build an analogy and derive from it original results. One of them was again the mentioned random-walk generating function [11], and the other the tight-binding Green function equation for electronic band structure on regular lattices [12]. Again, the two central expressions for two different physical systems had in common a convolution, and this was used to make the analogy concrete [I.ii.1, I.ii.2]. The known trapping procedure to determine first passage times [13] lead to electronic band structures for surfaces and thin films. To our knowledge there is no parallel version or development of this analogy. Again, independence of events, linearity and lattice regularity were (and are) the limitations of this approach.

I.ii.1 Surface electronic Green's functions in terms of the bulk Green's function. C. Varea, A. Robledo. *Phys. Rev. B* **19**, 1310 (1979) https://doi.org/10.1103/PhysRevB.19.1310

I.ii.2 Tight binding Green's functions for surfaces, thin films and solid interfaces. A random walk approach. A. Robledo, C. Varea. *Phys. Rev. B* **21**, 1469 (1980) https://doi.org/10.1103/PhysRevB.21.1469

I.iii First passage times for multiple walkers

This development [I.iii.1] anteceded similar studies [14]. Instead of the common setting of one random walker and one trap, we considered the successive trapping of an arbitrary number of independent walkers placed on a lattice each with different given initial position. The lattice contained also an arbitrary number of given sites that acted as irreversible traps. The literature at the time reported modeling the initial events of photosynthesis via trapping of random walkers. But the sites relevant for the triggering of the next, chemical, stage required the trapping of two walkers. Montroll's generating function method [8] was generalized to multiple walkers for both basic walks and continuous time walks [I.iii.1]. Independence of walkers, of walker steps, linearity and lattice regularity were (and are) the limitations of this approach.

I.iii.1 Multiple trapping of random walkers on periodic space lattices. A. Robledo, L. Woodhouse. *J. Stat. Phys.* **19**, 129 (1978) https://doi.org/10.1007/BF01012507

I.iv   Renormalization group and entropy for Weierstrass walks

Over the years we carried out our statistical-mechanics studies, which included the consideration of phase transitions, mainly for equilibrium states, and mainly under mean-field approximations. So, we were disconnected for a long time from the mainstream developments in critical phenomena and other topics involving scale invariance. When we finally considered the Renormalization Group (RG) method [I.iv.1] it was not with any specific application in mind, but instead we were interested in exploring its foundations. The RG method appeared to be a procedure that required devising a specific transformation for each application [2]. The RG transformation leads to flows away from nontrivial and towards trivial fixed points. This suggested the monotonic variation of a function akin to a thermodynamic potential, like entropy, as the nature of the system changed under the RG transformation. We looked for an optimization principle as the missing piece. To this end we selected the Weierstrass random walk [10] as a suitable model to investigate this proposition. The scaling properties of the single step distribution of this walk had been previously found to have similarities with the scaling of thermodynamic properties at criticality [10]. We generalized the Weierstrass walk to an infinite family of related walks with long-tailed single-step distributions such that the original walk was their RG nontrivial fixed point and corroborated that the entropy of the step distribution complied with functional optimization with the fixed points being entropy extrema. Years later we corroborated this significant feature of the RG method for network systems [III.viii.1, III.viii.2, III.viii.4, III.viii.5, III.viii.7]. To our knowledge there is no similar development of this property.

I.iv.1 Renormalization group, entropy optimization, nonextensivity at criticality. A. Robledo. *Phys. Rev. Lett.* **83**, 2289-2292 (1999) https://doi.org/10.1103/PhysRevLett.83.2289

I.iv.2 The renormalization group and optimization of entropy. A. Robledo. *J. Stat. Phys.* **100**, 475 (2000). https://doi.org/10.1023/A:1018620618862

I.iv.3 Anomalous diffusion, the renormalization group and optimization of entropy. A. Robledo, J. Quintana. *Granular Matter* **3**, 29-32 (2001). https://doi.org/10.1007/s100350000064

I.iv.4 Scale-invariant random walks and optimization of non-extensive entropy. A. Robledo, J. Quintana. *Chaos, Solitons and Fractals* **13**, 521-528 (2002) https://doi.org/10.1016/S0960-0779(01)00035-2

I.iv.5 "Unifying laws in multi-disciplinary power-law phenomena: fixed-point universality and non-extensive entropy". A. Robledo. In: Non-extensive Entropy-Interdisciplinary Applications, Oxford University Press, C. Tsallis and M. Gell-Mann, editors (2004) pp. 63-78. https://doi.org/10.1093/oso/9780195159769.003.0008

I.v Thermal-system statistical-mechanical analogy for renewal processes

We considered Montroll's [9] representation of the generating function approach for renewal processes and observed that this formalism is analogous to the statistical-mechanical description of non-interacting system with two fields, temperature and external field (or chemical potential) [I.iv.1, I.iv.2]. The generating function approach allows for the identification of four different ensembles and their partition functions (microcanonical, canonical, grand canonical, and the fourth ensemble that completes the set of conjugate variables for the Euler relation). Events and time represent degrees of freedom and energy, respectively. The analogy is straightforwardly extended to renewal processes with correlated events and, as expected, the concept of a waiting time (between events) distribution is no longer useful. But to each two-field thermal system corresponds a specific renewal event process. As a mathematical curiosity a renewal process consisting of q-independent events was analyzed, where the normal product of Laplace transformed waiting time distributions is replaced by the so-called q-product [I.iv.1].

I.v.1 "Renewal stochastic processes with correlated events. Phase transitions along time evolution". J. Velázquez, A. Robledo. *Phys. Rev. E* **83**, 031103-1-031103-6 (2011). https://doi.org/10.1103/PhysRevE.83.031103

I.v.2 "Statistical-mechanical structure for renewal stochastic processes". J. Velázquez, A. Robledo. *IJAMAS* **26**, 3-15 (2012). https://www.researchgate.net/publication/215439762

I.vi Phase transitions along time in correlated renewal processes

We made specific use of the above-referred renewal process approach that we had shown to be formally equivalent to the framework formed by the set of classical ensembles of equilibrium statistical mechanics [I.iv.2]. Most attention has been given to renewal processes of identical independently distributed (iid) events, but our approach can equally consider correlated events. Since each specific renewal event process can be translated into its equivalent (two-field) statistical-mechanical model system and vice versa, we chose a well-known solvable statistical-mechanical model exhibiting a continuous phase transition, the Hamiltonian Mean Field (HMF) model [15, 16], and we made use of it to describe a phase transition occurring along time. Recently, we made use of this exact analogy to develop an original stochastic approach for cascading identical events such as contagions in an epidemic or pandemic with application to the COVID 19 contingency. The cascade is made of time shifted renewal event chains such that its properties are derived from those of a renewal process, and as stated is in turn equivalent to a statistical-mechanical thermal system. We chose the HMF model and its phase transition to describe the effect of population confinement on the spread of contagions.

I.vi.1 "Renewal stochastic processes with correlated events. Phase transitions along time evolution". J. Velázquez, A. Robledo. *Phys. Rev. E* **83**, 031103-1-031103-6 (2011). https://doi.org/10.1103/PhysRevE.83.031103

I.vi.2 Statistical mechanical model for growth and spread of contagions under gauged population confinement. C. Velarde, A. Robledo. *Physica A* **573**, 125960 (2021). https://doi.org/10.1016/j.physa.2021.125960

II Density Functional Theory

Inhomogeneous states. Benjamin Widom considered the configurational (statistical mechanical) effect of the addition of a degree of freedom in an equilibrium system already composed of many of them [17]. He found a route, amenable for the study of inhomogeneous systems [18], to evaluate the equation of state for the chemical potential. We found that Widom's expression is the Euler-Lagrange equation (the vanishing of the first variation) associated with the corresponding free energy density functional. This identification opened many possible studies of equilibrium states, particularly non-uniform ones, like spatially ordered states, e.g., solids, phase coexistence, surfaces, etc. The additional degree of freedom could be a particle, a spin flip, different types of molecules in a mixture, even a step in a walk, an event in a renewal process, and so on.

Through this statistical-mechanical route we obtained descriptions for: liquid to solid transitions in hard core systems, like hard spheres; interfacial structures for the known (van der Waals) types of binary mixtures; kinetics of phase change, growth via interface advance, nucleation and spinodal decomposition. We determined global phase diagrams for wetting transitions at fluid mixture interfaces, and developed lattice models for micellar solutions, block copolymers and microemulsions. We also analyzed the modifications undergone by surface phase transitions through capillary confinement, including nematics and quiral molecule mixtures. In relation to wetting transitions we explored the features of a different type of inhomogeneity, that when interfaces meet and produces a line of tension. Finally, a particularly extensive set of studies contain the global phase diagram for the antiferromagnetic (mean field) spin-1 model, that complements the rich phase behavior of the (mean field) ferromagnetic case studied previously by Griffiths.

Another formal advance we carried out within density functional theory was to extend the customary approach to study a planar interface to a full three-dimensional inhomogeneity. This effort yielded statistical mechanical expressions for the rigidities of curved interfaces that are now added to their counterpart for the surface tension, the so-called Triezenberg-Zwanzig expression derived a couple of decades earlier. Furthermore, this development made contact with and gave microscopic support to the phenomenological free energy expression of Helfrich, a basic starting expression for the study of microemulsions. Also we extended our approach to the already mentioned line tension and in general to the stress tensor associated with deformations descriptive of more general inhomogeneous systems.

This second development led to an additional set of more specific results, amongst them the suggestion of a new kind of two-dimensional phase transitions, mediated by the interfacial curvatures of interfaces. For instance, the transformation of a singly-connected surface into either multiply-disconnected or multiply-connected surfaces. The interfacial rigidity appears as a relevant field in a free energy quantified via the genus of the surface and the Gauss-Bonnet theorem. It indicates a possible mechanism for the inception of micellar solutions or bicontinuous microemulsions from saturated amphiphile monolayers. Another case arises when considering the arrest, mediated by the presence of amphiphiles of a phase separation process, such as nucleation or spinodal decomposition. This stoppage offers another view of microemulsion formation. Finally, an extension of the capillary wave model for interfacial curvature fluctuations followed too.

A third stage was to focus on the basic procedure of the density functional approach. This consists of two steps, the vanishing of first variation of the free energy functional and the calculation of the sign of its second variation fixed by the solution of the 1st condition. The former locates the stationary solutions while the latter discriminates between maxima, minima or inflexion-point solutions. As it had been earlier noticed by Widom [19], the 1st condition is equivalent to a classical-mechanical particle problem, while the 2nd condition can be expressed as a quantum-mechanical particle eigenfunction problem. In our work these mechanical analogies were used to identify statistical-mechanical instabilities in simple liquids: the irrelevance for the position of the Gibbs dividing surface for planar interfaces, the Raleigh breakup of cylindrical jets, and the Laplace equation at the onset of spherical droplet nucleation. Other more complex behaviors were obtained for complex fluids containing amphiphiles.

Additionaly, we made use of our statistical-mechanical know-how in relation to the discovery of High T_c superconductivity and the frenzy of research activity that took place after this. From the body of accumulated experimental evidence a detailed picture was put together of a characteristic phase diagram for these materials. A rich assembly of properties that involves structural phase transitions driven by small variations of impurities or vacancies, that in turn triggers other phase changes, from antiferromagnetic to conducting, to superconducting phases. And to add interest, the mystery of the pairing mechanism. We contributed with a plausible complete description for the different phase behaviors and a novel proposition for a pairing mechanism based on the two-dimensional topological Kosterlitz-Thouless class of phase transitions [20].

Finally, as we describe below in Part IV, recently we have made use of density functional theory, in the form of a discrete-time Landau equation, to link this subject to low-dimensional nonlinear dynamics, their RG fixed-point maps for the routes to chaos, and the understanding of the generalized statistical mechanics represented by the Tsallis entropy expression.

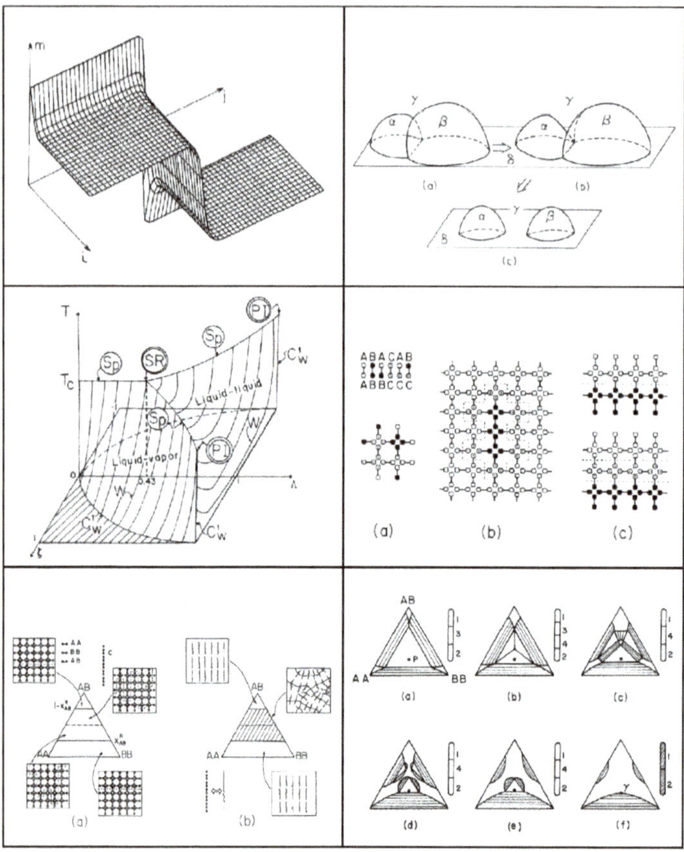

Figure 2: Wetting phenomena and complex fluid models. The top left panel shows the density or composition inhomogeneity that becomes a line of tension in macroscopic scales. The top right panel illustrates drops of different liquid phases on a flat solid surface that form lines of tension when three different interfaces meet. The left middle panel shows a global phase diagram for wetting and prewetting interfacial transitions and where different types of multicritical points occur. Sections II.iv and II.x refer to studies related to these figure panels. The right middle panel illustrates lattice model configurations for different types of bifunctional molecules that are placed along lattice bonds with certain rules. The bottom panels show phase diagrams for complex fluids obtained for these types of lattice models. Section II.v refers to studies related to these figure panels.

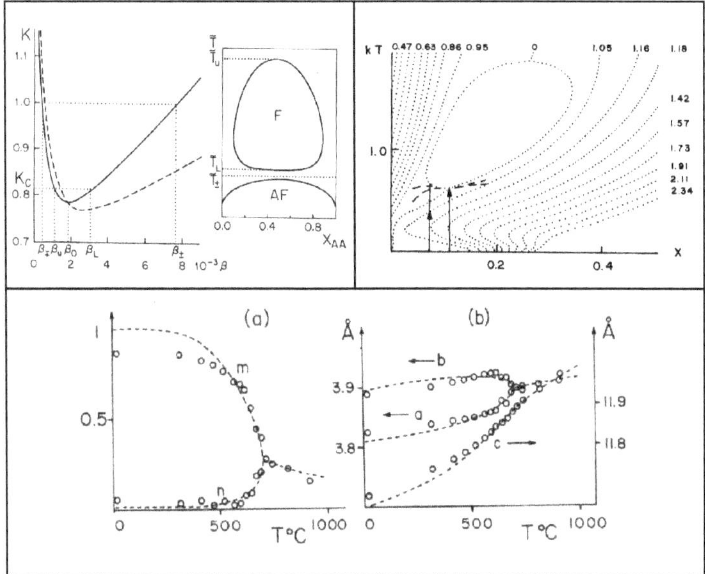

Figure 3: Reentrant solubility and High-T_c superconductors phase behavior. The top panels show the phenomenon of reentrant solubility that involves two different critical miscibility points. The plots correspond to micellar solution lattice models that map into the Ising magnet model. The asymmetric case involves anomalous criticality. Section II.vi refers to studies related to these figure panels. The bottom panels show model phase diagrams for sublattice ordering and structural transitions and comparison with experimental counterpart in High-T_c copper-oxide superconductors. Section II.xii refers to studies related to these figure panels.

II.i Density functional theory and Widom's particle insertion method

In the late Seventies interest appeared in the study of inhomogeneous equilibrium states [21, 22], such as fluid-wall and fluid-fluid interfaces, and the phase transition from a fluid to a solid [23]. Two approaches emerged, density functional theory [21, 22] and Widom's particle insertion method [17, 18]. Our early interests focused on exhibiting the basic link between them. It was shown that Widom's particle insertion formula is equivalent to the Euler-Lagrange equation, the vanishing of the first variation of the free energy density [II.i.1]. To our knowledge there is no other development of this equivalency. Widom's formula and the corresponding density functional were then determined for the Kac potential in the long range, van der Waals limit (a Hamiltonian route for mean field approximation) and applied to obtain global properties for binary fluid mixtures [II.i.2–II.i.4].

II.i.1 On the relationship between the density functional formalism and the potential distribution theory for nonuniform systems. A. Robledo, C. Varea. *J. Stat. Phys.* **26**, 513 (1981) https://doi.org/10.1007/BF01011432

II.i.2 Rigorous interface properties of the van der Waals mixture. A. Robledo, A. Valderrama, C. Varea. *J. Chem. Phys.* **73**, 6365 (1980) https://doi.org/10.1063/1.440123

II.i.3 Free energy density functionals for nonuniform classical fluids. In: Keller, J., Gázquez, J.L. (eds) Density Functional Theory. A. Robledo, C. Varea. *Lecture Notes in Physics* **187**, 287 (1983) https://doi.org/10.1007/3-540-12721-6_10

II.i.4 Global behavior of the diffusion coefficient for the van der Waals binary mixture. R. Castillo, M.E. Costas, A. Robledo. *Int. J. Thermophys.* **10**, 427 (1989) https://doi.org/10.1007/BF01133539

II.ii Liquid to solid transitions of hard-core model systems

Also in the late Seventies, in an attempt to 'derive' from first principles the liquid-solid transition in hard-core fluids we looked first at the simplest possible model, a lattice gas with nearest-neighbor exclusion, and applied to it a method (our own implementation of Widom's formula) that many years later we realized was exact for a Bethe lattice, that is, the mean field approximation! in a version suitable for hard core interactions. The next stage was to solve the problem for an extended hard-core up to an unspecified number of nth-order neighbors. The third stage was to obtain the continuum space limit by allowing n go to infinity while the lattice spacing went to zero. The result is a family of Bethe lattice models that exhibit 'liquid to solid' transitions and that are solved exactly. The main finding for hard spheres is the occurrence of a bifurcation from the uniform liquid solution at (random) close packing to a periodic solid-like density. Following the periodic solution from the bifurcation decreases its density until it can reverse its course and increase. A density can be determined such that there is a first order transition before close packing. Thus, the uniform liquid branch turns metastable at the liquid-solid transition. See Ref. [II.ii.1]. Our derivation of the exact solutions for correlations of the finite system of hard rods appeared in Ref. [II.ii.2]. Years later the family of Bethe lattice hard-core exclusion models was revisited and their properties described in further detail [II.ii.3]. There have been many density functional studies of the liquid-solid transition for hard sphere systems over the years but they do not encounter or consider the bifurcation at random or uniform close packing [24].

II.ii.1 The liquid-solid transition of the hard sphere system from uniformity of the chemical potential. A. Robledo. *J. Chem. Phys.* **72**, 1701 (1980) https://doi.org/10.1063/1.439281

II.ii.2 The distribution of hard rods on a line of finite length. A. Robledo, J.S. Rowlinson. *Mol. Phys.* **58**, 711 (1986) https://doi.org/10.1080/00268978600101521

II.ii.3 The hard-sphere order-disorder transition in the Bethe continuum. A. Robledo, C. Varea. *J. Stat. Phys.* **63**, 1163 (1991) https://doi.org/10.1007/BF01030004

II.iii   Kinetics of phase change, nucleation, spinodal decomposition

The Landau and the Cahn-Hilliard kinetic equations [25, 26, 27], together with the Metiu-Kitahara-Ross extension [28, 29, 30], had been previously applied to describe kinetics of phase change in fluid models under the widely employed square-gradient approximation (the Landau density functional suitable for slowly varying conditions near critical points). Our intervention on this topic involved the use of a more general density functional that we obtained from our experience with Widom's insertion method. This functional introduces spatial non-locality and revealed [II.iii.1] previously unknown features, such as: i) Cooperation of non-locally correlated fluctuations that provide the nucleation process with an additional term absent in the classical and gradient theories. ii) And an infinite family of periodic stationary states within the spinodal region that serve as a ladder to reach stability in the approach to equilibrium. Our advance opened the way for: the determination of similar features in segregating and ordering alloys [II.iii.2]; an extension of irreversible thermodynamics [II.iii.3]; the observation of the effect of confinement on spinodal decomposition [II.iii.4]; and the influence of wetting on kinetics of phase change in fluid mixtures [II.iii.5].

II.iii.1 Nucleation, spinodal decomposition and kinetics of phase change in the van der Waals fluid. C. Varea, A. Robledo. *J. Chem. Phys.* **75**, 5080 (1981) https://doi.org/10.1063/1.441899

II.iii.2 Kinetics of phase change in a model binary alloy. A. Robledo, C. Varea. *Phys. Rev. B* **25**, 4711, (1982) https://doi.org/10.1103/PhysRevB.25.4711

II.iii.3 A nonlocal theory of irreversible thermodynamics. A. Robledo, C. Varea. *J. Noneq. Thermodyn.* **12**, 213 (1987) https://doi.org/10.1515/jnet.1987.12.3.213

II.iii.4 Spinodal decomposition under confinement. C. Varea, J. Campos-Terán, A. Robledo. *Physica A* **244**, 440-452 (1997). https://doi.org/10.1016/S0378-4371(97)00228-8

II.iii.5 Kinetics of phase change in binary mixtures with complete wetting interfaces. J. Quintana, A. Robledo. *Mol. Phys.* **95**, 587-593 (1998) https://doi.org/10.1080/00268979809483192

II.iv   Global wetting phase diagram for fluid interfaces

Our contributions were not the earliest wetting transition developments, mostly done for planar solid to fluid interfaces [31, 32, 33, 34, 35], but the first dealing with fluid-fluid interfaces. We constructed a global phase diagram for the entire set of interfaces of the van der Waals binary fluid mixture types [II.iv.1] that is the counterpart for the global phase diagram for planar solid to fluid interfaces [36]. The methodology was mostly based on a non-local density functional that generalized the common slowly varying square gradient approximation. We addressed some unusual systems that did not receive attention in other studies of wetting, such as the wetting properties of model semipermeable membranes [II.iv.3, II.iv.4], pinning of alloy anti-phase boundaries [II.iv.5], and the effect of wetting on the kinetics of phase change [II.iv.8]. We also studied experimentally some related phenomena, such as the possible presence of prewetting in transient foams [II.iv.6], and of a continuous wetting transition in a specific fluid mixture [II.iv.7].

II.iv.1 Global phase diagram for the wetting transition at interfaces in fluid mixtures. M.E. Costas, C. Varea, A. Robledo. *Phys. Rev. Lett.* **51**, 2394 (1983) https://doi.org/10.1103/PhysRevLett.51.2394

II.iv.2 Prewetting in partially miscible liquids and the structure and thermodynamics of transient foams and aerosols. J. Gracia, C. Varea, A. Robledo. *J. Phys. Chem. (Lett.)* **88**, 3923 (1984) https://doi.org/10.1021/j150662a004

II.iv.3 Wetting regimes at semipermeable membranes. C. Varea, A. Robledo, E. Martina. *Phys. Rev. A* **31**, 1825 (1985) https://doi.org/10.1103/PhysRevA.31.1825

II.iv.4 Interfacial critical phenomena at semipermeable membranes. A. Robledo, C. Varea, E. Martina. *Phys. Rev. B* **32**, 7545 (1985) https://doi.org/10.1103/PhysRevB.32.7545

II.iv.5 Pinning of antiphase boundaries at the cleaved (001) surface of an $L2_0$ ordering alloy. L. Vicente, C. Varea, A. Robledo. *Surf. Sci.* **164**, 479 (1985) https://doi.org/10.1016/0039-6028(85)90761-7

II.iv.6 Transient foaminess, micelle formation and wetting behavior in water-phenol mixtures. J. Gracia, C. Guerrero, J. Llañes, A. Robledo. *J. Phys. Chem.* **90**, 1350 (1986) https://doi.org/10.1021/j100398a028

II.iv.7 Continuous wetting transition at the liquid-vapor interface of the binary liquid mixture cyclohexane-acetonitrile. L.M. Trejo, J. Gracia, C. Varea, A. Robledo. *EPL* **7**, 537 (1988) https://doi.org/10.1209/0295-5075/7/6/010

II.iv.8 Kinetics of phase change in binary mixtures with complete wetting interfaces. J. Quintana, A. Robledo. *Mol. Phys.* **95**, 587-593 (1998) https://doi.org/10.1080/00268979809483192

II.v Lattice models for micellar solutions, microemulsions, magnetic alloys, etc.

We made an attempt to contend with the teams seeking to develop a lattice model to describe the interesting microemulsion phase, and the transitions it undergoes. Generalizations of the original Widom-Wheeler model [37, 38] were developed to address different types of systems. The first of them was designed to be equivalent to the solid-on-solid Ising model that was found to display the elusive roughening transition [39]. Through an exact analogy a representation of the disordering of a lamellar diblock polymer microemulsion was obtained [II.v.1]. A second generalization [II.v.2] to an additional type of molecular end (of the two original types) that form the bifunctional molecules, the Widom-Wheeler mixture model expanded in such a way as to bc exactly analogous to the famous Griffiths' three-component model (the mean-field spin-1 model) [40]. A third generalization [II.v.3–II.v.6] to finite interactions of the original Widom-Wheeler model led not only to a rich global micellar solution model phase behavior, but also (again but in a different way) to an analog of the symmetric portion of the Griffiths three-component model with staggered field [II.v.5]. Finally, we extended Griffiths original work, a vast global phase diagram for uniform, or ferromagnetic, phases to sublattice ordered, or antiferromagnetic, phases [II.v.7]. New applications appeared for ordering magnetic alloys [II.v.8] and to micellar solution models [II.v.9]. The most important contribution [II.v.7] was revealing a rich phase behavior where ordered phases are present that had remained hidden under the known global phase diagram for uniform phases made by Griffiths and coworkers. The global phase diagrams in Refs. [40] and [II.v.7] are like the two different faces of the moon.

II.v.1 Roughening transition and formation of bicontinuous structures of immiscible solvents embedded in surfactant diblock copolymers. A. Robledo, C. Varea, Martina, E.. *J. Phys. Lett. (fr)* **46**, L-967 (1985) https://doi.org/10.1051/jphyslet:0198500460 20096700

II.v.2 Relationships between the phase behavior of lattice models of amphiphile mixtures and Griffiths's Three-Component Model. C. Varea, A. Robledo. *Phys. Rev. A* **33**, 2760 (1986) https://doi.org/10.1103/PhysRevA.33.2760

II.v.3 Exact thermodynamic correspondence between a lattice model microemulsion and simpler spin systems . A. Robledo. *EPL* **1**, 303 (1986) http://iopscience.iop.org/0295-5075/1/6/006

II.v.4 Critical magnetization at antiphase boundaries of magnetic binary alloys. C. Varea, A. Robledo. *Phys. Rev. B* **36**, 5561 (1987) https://doi.org/10.1103/PhysRevB.36.5561

II.v.5 Spin Ising transcription of a lattice model of micellar solutions. A. Robledo. *Phys. Rev. A* **36**, 4067 (1987) https://doi.org/10.1103/PhysRevA.36.4067

II.v.6 Statistical mechanical models for micellar solutions and microemulsions. A. Robledo. Proceedings of the Fourth Mexican School on Statistical Physics, World Scientific, Singapore, 1988, pp. 93-165

II.v.7 Sublattice ordered phases of the Griffiths' Three-Component model. V. Talanquer, C. Varea, A. Robledo. *Phys. Rev. B* **39**, 7016 (1989) https://doi.org/10.1103/PhysRevB.39.7016

II.v.8 Global phase diagram for binary alloys with one magnetic component. V. Talanquer, C. Varea, A. Robledo. *Phys. Rev. B* **39**, 7030 (1989) https://doi.org/10.1103/PhysRevB.39.7030

II.v.9 Sublattice-ordered phases in a model for a micellar solution. V. Talanquer, C. Varea, A. Robledo. *Phys. Rev. B* **39**, 7039 (1989) https://doi.org/10.1103/PhysRevB.39.7039

II.vi Anomalous micellar solubility loops

Nonionic amphiphile aqueous solutions exhibit asymmetric solubility loops and experimental determination of critical exponents at their lower critical solubility point revealed anomalous behavior suggesting nonuniversality of the critical indexes [41, 42]. The extension of the Widom-Wheeler model we had made to finite interactions and its analogue as a micellar solution model became a suitable platform to obtain, first of all, solubility loops [II.vi.1, II.vi.2], and then, the more interesting and experimentally available, asymmetric solubility loops [II.vi.3, II.vi.4]. We therefore had the opportunity to investigate theoretically the experimentally detected critical behavior and address the polemical observations. The model is equivalent to a temperature and external field dependent spin-1/2 Ising magnet, and this feature provides a rationalization that explains the controversial experimental observations, that in turn led to explicit approbatory reference [43]. Our model was suitable for further studies such as micellar interfacial and capillary properties [II.vi.5] that resulted in the characterization of enhanced amphiphile adsorption at their liquid-liquid interfaces [II.vi.6].

II.vi.1 New source of corrections to scaling for micellar solution critical behavior. G. Martinez-Mekler, G.F. Al-Noaimi, A. Robledo. NATO Advanced Study Institute Series B, Pergamon Press, 1989, pp. 211-215

II.vi.2 Uncommon source of corrections to scaling for micellar solution critical behavior. G. Martinez-Mekler, G.F. Al-Noaimi, A. Robledo. *Phys. Rev. A* **41**, 4513 (1990) https://doi.org/10.1103/PhysRevA.41.4513

II.vi.3 Micellar solution model with asymmetric solubility loops and anomalous critical isochore with crossover from angular to parallel approach to coexistence. A. Robledo, G. Martinez-Mekler, C. Varea. Lecture Notes in Thermodynamics and Statistical Mechanics M. Lopez de Haro and C. Varea, eds. p. 14, World Scientific, 1990.

II.vi.4 Asymmetric solubility loops and anomalous geometry at the lower critical point in a model micellar solution. A. Robledo, G. Martinez-Mekler, C. Varea. *EPL* **16**, 405 (1991) https://doi.org/10.1209/0295-5075/16/4/015

II.vi.5 Interfacial and capillary properties of a micellar solution model. C. Varea, C. García-Alcántara, A. Robledo. *Physica A* **236**, 177-187 (1997). https://doi.org/10.1016/S0378-4371(96)00397-4

II.vi.6 Enhanced amphiphile adsorption at liquid-liquid interfaces in a micellar solution model. C. García-Alcántara, C. Varea, A. Robledo. *Physica A* **256**, 321-332 (1998) https://doi.org/10.1016/S0378-4371(98)00196-4

II.vii Complex fluids under confinement

Density functional theory proved to be a suitable method for the study of the finite size effects on the properties of inhomogeneous systems. Phase transitions are modified when a three-dimensional system is confined in size along only one spatial dimension [44]. Some years after the topic received the first wave of attention by the statistical physics community we analyzed a few new situations. Some of them correspond to the same model studies that ante-ceded us but aimed at gaining new knowledge from them. We studied the modification undergone by spinodal decomposition under confinement and observed the development of lamellar or strip patterns [II.vii.1]. Other works aimed at the description of phase transitions induced or suppressed by confinement [II.vii.6] that become relevant when the models represent nematic liquids [II.vii.2] or enantiomeric mixtures [II.vii.5]. Additionaly, we put together a global description for the effects of confinement through the family of surface phase transitions exhibited by a semi-infinite model magnet or simple fluid [II.vii.3, II.vii.7]. We also determined density fluctuations and correlations when a model system becomes finite along one dimension [II.vii.4].

II.vii.1 Spinodal decomposition under confinement. C. Varea, J. Campos-Terán, A. Robledo. *Physica A* **244**, 440-452 (1997). https://doi.org/10.1016/S0378-4371(97)00228-8

II.vii.2 Phase properties of nematics confined by competing walls. J. Quintana, A. Robledo. *Physica A* **248**, 28-43 (1998). https://doi.org/10.1016/S0378-4371(97)00523-2

II.vii.3 Landau density functional theory for one-dimensional inhomogeneities. A. Robledo, J. Quintana. *Physica A* **257**, 197-206 (1998) https://doi.org/10.1016/S0378-4371(98)00140-X

II.vii.4 Density fluctuations and correlations of confined fluids. C. Varea, A. Robledo. *Physica A* **268**, 391-411 (1999) https://doi.org/10.1016/S0378-4371(99)00049-7

II.vii.5 Confinement induced immiscibility of mixtures of enantiomers. J. Quintana, A. Robledo. *Physica A* **295**, 333-347 (2001). https://doi.org/10.1016/S0378-4371(01)00129-7

II.vii.6 "Phase transitions induced or suppressed by confinement". A. Robledo, J. Quintana. NATO Advanced Research Series: "New kinds of phase transitions: transformations in disordered substances", Kluwer Academic Publishers 2002, pp. 545-555.

II.vii.7 "Surface transitions under confinement". J. Quintana, A. Robledo. *J. Phys.: Cond. Matt.* **14**, 2211-2221 (2002). https://doi.org/10.1088/0953-8984/14/9/310

II.viii   Curved interfaces, bending rigidities, line tension, stress tensor and capillary waves

We considered density functional theory for inhomogeneous systems for which surface tension is not the only important or the main free energy contribution, and determined bending constants, interfacial width and line tension [II.viii.1]. In relation to the microemulsion problem we went smoothly from lattice models, and studied the properties of curved interfaces in continuum space, of particular importance because of the vanishing of the surface tension. From a 'microscopic' starting point we derived expressions for the bending terms that are 'mesoscopic' quantities, going one step beyond the usual surface tension description. Indeed, by considering cylindrical, spherical, and then general surface shape fluctuations it was possible to derive rigorous expressions for the bending constants in terms of pair correlations [II.viii.2, II.viii.3], similar to the expression of Triezenberg and Zwanzig for the surface tension obtained many years before. Amongst other results we, and collaborators, were able to generalize the capillary-wave model of an interface and explain its properties [II.viii.4, II.viii.7, II.viii.11, II.viii.12]. We also derived explicit expressions for the stress tensor for general inhomogeneities in a one-component simple fluid in terms of density gradients and moments of the direct correlation function [II.viii.5, II.viii.9]. We were the first with these closed-form expressions, and with the conceptual clarification of the phenomenological Helfrich free energy that governs the physics of curved interfaces [II.viii.6, II.viii.8, II.viii.10, II.viii.13]. Our results contributed to provide the statistical-mechanical basis for the free energy terms of curved interfaces originally derived from phenomenological elasticity arguments.

II.viii.1 Density functional theory beyond the surface tension. Interfacial width, elastic bending constants and line tension. A. Robledo, C. Varea, V. Romero-Rochín. *Physica A* **177**, 474 (1991) https://doi.org/10.1016/0378-4371(91)90189-J

II.viii.2 Microscopic expressions for interfacial bending constants and spontaneous curvature. V. Romero-Rochín, C. Varea, A. Robledo. *Phys. Rev. A* **44**, 8417 (1991) https://doi.org/10.1103/PhysRevA.44.8417

II.viii.3 Bending rigidity of the liquid-vapor interface. V. Romero-Rochín, C. Varea, A. Robledo. Lectures on Thermodynamics and Statistical Mechanics, World Scientific, Singapore, pp. 10-25, 1991.

II.viii.4 Extended capillary-wave theory for the liquid-vapor interface and its width in the limit of vanishing gravity. V. Romero-Rochín, C. Varea, A. Robledo. *Physica A* **184**, 367 (1992) https://doi.org/10.1016/0378-4371(92)90312-E

II.viii.5 Stress tensor of curved interfaces. V. Romero-Rochín, C. Varea, A. Robledo. *Mol. Phys.* **80**, 821 (1993) https:/doi.org/10.1080/00268979300102681

II.viii.6 Free energy expressions for a spherical interface. C. Varea, A. Robledo. *Mol. Phys.* **85**, 477 (1995). https://doi.org/10.1080/00268979500101261

II.viii.7 Scaling properties of the capillary-wave model with interfacial bending rigidity. A. Robledo, C. Varea. *Mol. Phys.* **86**, 879 (1995) https://doi.org/10.1080/0026897950 0102451

II.viii.8 Can the Helfrich free energy for curved interfaces be derived from first principles?. A. Robledo, C. Varea. *Physica A* **231**, 178-190 (1996). https://doi.org/10.1016/ 0378-4371(95)00457-2

II.viii.9 Stress tensor of inhomogeneous fluids. C. Varea, A. Robledo. *Physica A* **233**, 132-144 (1996). https://doi.org/10.1016/S0378-4371(96)00244-0

II.viii.10 Scaling of interfacial tension and identity of bending moduli of microemulsions. A. Robledo, C. Varea. International School of Physics "Enrico Fermi", Course CXXXIV, "The Physics of Complex Fluids", julio 9-19 1996, Varenna, Italia, IOS Press 1997, pp 417-431

II.viii.11 Interfacial width and shape fluctuations and extensions of the gaussian model of capillary waves. A. Robledo, C. Varea. *J. Stat. Phys.* **89**, 273-282 (1997). https://doi.org/10.1007/BF02770765

II.viii.12 Fluctuations and instabilities of model amphiphile interfaces. C. Varea, A. Robledo. *Physica A* **290**, 360-378 (2001). https://doi.org/10.1016/S0378-4371(00)00461-1

II.viii.13 Theory of interfacial bending constants. C. Varea, A. Robledo. *J. Phys.: Cond. Matt.* **13**, 9075-9088 (2001). https://doi.org/10.1088/0953-8984/13/41/303

II.ix Curvature interfacial transitions

As a consequence of our understanding of the Helfrich free energy of curved interfaces we were in a position to suggest the possibility of novel types of surface phase transitions that could take place in thin interfaces like such as amphiphile interfaces. The first new kind of transition corresponds to multi-patch buckling of the surface and its associated order parameter is its mean curvature [II.ix.1–II.ix.3]. The second, topologically-driven, novel transition transforms a simply-connected surface state into a multiply disconnected (or multiply connected) volume-filling surface state. Its order parameter is given by the surface genus (or number of micelles). These volume-filling surface states arise when the free energy cost for the creation of a droplet or a 'handle' out of the original monolayer surface vanishes. Our results follow from the Gauss-Bonnet theorem that links local curvature properties with the surface global topological invariants [II.ix.1–II.ix.3]. Finally, we made use of the results from the Helfrich free energy studies to obtain an interpretation of the phase properties of microemulsions as a detained, or arrested, ordinary phase separation due to the action of amphiphiles [II.ix.4]. That is, we put forward an interpretation of the characteristic phase properties of microemulsions based on a distinct process of phase separation that has come to a stand-still due to the action of the amphiphiles on the driving force of the process. In our kinetic equation this force relates to the generalized Laplace equation that contains spontaneous curvature and bending rigidity terms, and, according to it, the final stationary states attained are phases structured into water-rich and oil-rich domains [II.ix.4].

II.ix.1 Curvature interfacial transitions at amphiphile monolayers and their possible relation to the onset of micelle formation. A. Robledo, C. Varea, V. Talanquer. Lecture Notes in Thermodynamics and Statistical Mechanics M. Lopez de Haro and C. Varea, eds. p. 3, World Scientific, 1990.

II.ix.2 Curvature interfacial transitions in amphiphile monolayers and their possible relation to the onset of micelle formation. A. Robledo, C. Varea, V. Talanquer. *Phys. Rev. A* **43**, 5736 (1991). https://doi.org/10.1103/PhysRevA.43.5736

II.ix.3 Interfacial phase transitions underlying amphiphile micellar self-assembly. A. Robledo, C. Varea. Lectures on Thermodynamics and Statistical Mechanics, World Scientific, Singapore, pp. 37-43, 1992.

II.ix.4 Arrested phase separation and phase equilibrium properties of microemulsions. A. Robledo, C. Varea. "International Workshop on the Morphology and Kinetics of Phase Separating Complex Fluids", Messina, Italia, junio 24-28 1997. Il Nuovo Cimento 20 D, 2315-2324 (1998)

II.x Line tension & wetting

In parallel with these developments we looked at a different type of inhomogeneity, the macroscopic line that is formed by the intersection of three interfaces and that carries a free energy cost, the line of tension. First: we considered the change in grand potential due to a line inhomogeneity where two or more interfaces meet, and derive an exact expression for its tension in terms of the direct correlation function and the gradients of the densities. We proposed a model where a line inhomogeneity is formed at the intersection of the free surface and a domain boundary of an Ising magnet [II.x.1, II.x.4]. Second: we calculated, for a spin-1 Ising model within the mean-field approximation, the line tension along partial-wetting surface states up to a first-order wetting transition where the line disappears. We likewise calculated the line tension of the boundary between the two coexisting surface states at the prewetting transition and follow its behavior into the neighborhood of bulk wetting. In both cases we find evidence for the divergence of the line tension [II.x.2]. Third: we extended the interfacial wetting transition to a similar phenomenon for the contact line together with an Antonov rule for the line tension [II.x.3]. The standard wetting transition consists of the transformation of a microscopically thin two-dimensional interface into a macroscopically thick structure composed of two interfaces separated by a bulk phase. We considered the one-dimensional analog of this phenomenon, when a contact line among three or more phases decomposes into two contact lines separated by an interface. Our findings help both in settling the discussion on the limiting value of the line tension and in understanding the origin of its singular behavior. Fourth: we studied the effect of fluctuations on the line tension at first-order and continuous wetting transitions. We considered thermal wandering, strong fluctuations as in random media, and weak fluctuations as in quasiperiodic systems. These properties establish the occurrence of hyperscaling and nonclassical exponents for the line tension at wetting. [II.x.6, II.x.8]. Fifth: we studied the three-phase contact line and its tension near the interfacial phase transition from partial to complete wetting. We analyzed the singularity of the line tension at first-order wetting transitions and showed that it displays the universal features of a critical endpoint [II.x.7].

II.x.1 Statistical mechanics of the line tension. C. Varea, A. Robledo. Lectures on Thermodynamics and Statistical Mechanics, World Scientific, Singapore, pp. 190-203, 1991.

II.x.2 Evidence for the divergence of the line tension at the wetting transition. C. Varea, A. Robledo. *Phys. Rev. A* **45**, 2645 (1992) https://doi.org/10.1103/PhysRevA.45.2645

II.x.3 Wetting transition for the contact line and Antonov's rule for the line tension. A. Robledo, C. Varea, J.O. Indekeu. *Phys. Rev. A* **45**, 2423 (1992) https://doi.org/10.1103/PhysRevA.45.2423

II.x.4 Statistical Mechanics of the line tension. C. Varea, A. Robledo. *Physica A* **183**, 12 (1992) https://doi.org/10.1016/0378-4371(92)90175-P

II.x.5 Magnitude of the prewetting boundary tension near wetting for short-range forces. C. Varea, A. Robledo. *Phys. Rev. E* **47**, 3772 (1993) https://doi.org/10.1103/PhysRevE.47.3772

II.x.6 Hyperscaling and nonclassical exponents for the line tension at wetting. J.O. Indekeu, A. Robledo. *Phys. Rev. E* **47**, 4607 (1993) https://doi.org/10.1103/PhysRevE.47.4607

II.x.7 Universality and the contact line at first-order wetting transitions. A. Robledo, J.O. Indekeu. *EPL* **25**, 17 (1994). https://doi.org/10.1209/0295-5075/25/1/004

II.x.8 Effect of thermal fluctuations on the singular behaviour of the line tension at wetting. J.O. Indekeu, A. Robledo. Proceedings of "Recent advances in Statistical Physics", *Turkish J. Phys.* **18**, 285 (1994).

II.xi   Analogy of density functional 1st & 2nd variations with classical and quantum mechanics

We studied first the stability of cylindrical and spherical interfaces with respect to density fluctuations within the square-gradient approximation [II.xi.1]. That is, we determined the stability matrix (of the second derivatives of the free energy functional with respect to the density) when the stationary state is a cylindrical or spherical droplet of a stable phase embedded in the metastable phase. For these geometries the stationary states are unstable, some of the eigenvalues are negative and their eigenfunctions represent those fluctuations that are amplified in a process where the equilibrium state is reached. At early times, in a simple Landau kinetics model, the eigenfunctions represent the fluctuations that grow or decay with a simple exponential law and with a characteristic time that is proportional to the inverse of the eigenvalues. Our results agree with the stability criteria obtained from the Laplace equation, that is, the nucleation of critical droplets and in the case of cylinders also the Rayleigh instability. In the limit of infinite radius we recover the known results for the planar interface between two stable phases. The modes with lowest energy correspond to the customary capillary waves, while other modes with higher energy associated to changes in the interfacial width are shown to be related to a novel interfacial coefficient. Subsequently, we carried out a comprehensive description of interfaces containing amphiphiles developed through the use of a free energy density functional with squared-gradient and squared-Laplacian terms [II.xi.2, II.xi.3]. This elemental model functional contains the basic ingredients to examine interfacial stability, it is technically tractable and a range of results for curved interfaces, many in analytical form, have been obtained from it. These are: (i) average equilibrium properties, such as pressure tensor, interfacial tension and position of the Gibbs dividing surface, (ii) order parameter fluctuation modes and stability matrix, and, (iii) an effective interfacial potential that in the small curvature limit corresponds to the Helfrich free energy. Our survey was meant to fill an existing gap, mostly conceptual, left by previous work.

II.xi.1 Fluctuations and instabilities of curved interfaces. C. Varea, A. Robledo. *Physica A* **255**, 269-284 (1998) https://www.academia.edu/52441404

II.xi.2 Fluctuations and instabilities of model amphiphile interfaces. C. Varea, A. Robledo. *Physica A* **290**, 360-378 (2001). https://doi.org/10.1016/S0378-4371(00)00461-1

II.xi.3 "Stability of curved amphiphilic interfaces". C. Varea, A. Robledo. *Physica A* **306**, 301-315 (2002). https://doi.org/10.1016/S0378-4371(02)00507-1

II.xii Phase behavior and pairing mechanism for two-dimensional superconductors

We studied the properties of the copper oxide High-T_c superconductors from a statistical-mechanical perspective. First, we developed a model for twin boundaries, like those observed in YBCO ceramic superconductors. The model, based on an inhomogeneous Landau-Ginsburg free energy, was used to show that the twin boundaries induce a purely interfacial superconducting transition at a temperature above the bulk transition temperature [II.xii.1, II.xii.3]. Second, we developed a model for the absorption of oxygen in the copper-oxide basal planes of YBCO ceramic superconductors based on a layered oxygen lattice gas model in equilibrium with an external oxygen source. We considered, in addition to Cu-mediated oxygen-oxygen interactions, the elastic energy of the crystal. The model exhibits coupled oxygen ordering and tetragonal to orthorhombic structural transitions. We obtained quantitative agreement with experimental data [II.xii.2, II.xii.4, II.xii.5]. Third, we developed a spin-hole model for the magnetic phase behavior of weakly coupled copper-oxygen layers. For the hole-free system there are antiferromagnetic couplings between Cu magnetic moments and weak Ising anisotropy with canting. Addition of holes, localized on the O sites, induces a transformation of anisotropy in the spin couplings, from Ising to XY, and of its sign, from antiferromagnetic to ferromagnetic [II.xii.6, II.xii.7]. Fourth, the spin-hole model was developed further via the use of the Moriya-Dzyaloshinsky Hamiltonian for anisotropic Heisenberg spin couplings. The global phase diagram for the ceramic superconductors became better rationalized. Addition of holes (via doping) smooths corrugation of the copper-oxygen planes, that in turn drives the structural phase transition from tetragonal to orthogonal, and changes the magnetic behavior from Ising-like antiferromagnetic with a sharp Neel temperature to XY-like with weak ferromagnetic couplings [II.xii.8, II.xii.9]. Fifth, the spin-hole model implies a novel kind of charge pairing mechanism for the copper oxide High Tc superconductors. The vortex magnetic excitations occurring in the copper-oxide planes capture mobile charges or holes, and the pairing of these occurs via the Kosterlitz-Thouless (KT) vortex-antivortex pairing mechanism.

[II.xii.7–II.xii.9]. Sixth, the phase properties of the spin-hole model were found to be consistent with an extension of the Hubbard Hamiltonian model to competing positive- and negative-U interactions on a 2D lattice. The positive-U Hubbard model exhibits Ising antiferromagnetic phase behavior, while the negative-U Hubbard model displays XY superconductivity. The phase progression observed in High-T_c superconductors, and in the spin-hole model, can be obtained via gradual change in the concentration of initially purely $U > 0$ interactions via addition of $U < 0$ interactions, like in a binary alloy [II.xii.10]. It is only recently that the Moriya-Dzyaloshinsky interaction and the KT transition has been associated with the properties of copper oxide High-T_c superconductors [45, 46].

II.xii.1 Oxygen ordering and twin boundaries in a Landau-Ginsburg superconductor oxide model. C. Varea, A. Robledo. "Novel Superconductivity", S.A. Wolf and V.Z. Kresin, Eds., Plenum, 1987, pp. 1033-1039

II.xii.2 Effect of oxygen pressure and temperature on the tetragonal-orthorhombic transition in a model $YBa_2Cu_3O_{7-y}$. C. Varea, A. Robledo. *Rev. Mex. Fís.* **33**, 311 (1987) https://rmf.smf.mx/ojs/index.php/rmf/article/view/1936

II.xii.3 High-Tc superconductivity at twin boundaries in a Landau-Ginzburg oxide model. A. Robledo, C. Varea. *Phys. Rev. B* **37**, 637 (1988) https://doi.org/10.1103/PhysRevB.37.631

II.xii.4 Oxygen absorption and structural transition in a model $YBa_2Cu_3O_{7-y}$. C. Varea, A. Robledo. *MRS Symp. Proc.* **99**, 527 (1988) https://doi.org/10.1557/PROC-99-527

II.xii.5 Model for oxygen absorption and structural phase transition in $YBa_2Cu_3O_{7-y}$. C. Varea, A. Robledo. *Mod. Phys. Lett. B* **2**, 1017, (1988) https://doi.org/10.1142/S0217984988000849

II.xii.6 Simple spin-hole model for magnetic correlations in copper-oxide superconductors. A. Robledo, C. Varea. *Int. J. Mod. Phys. B* **1**, 763 (1988) https://doi.org/10.1142/S0217979288000597

II.xii.7 Spin-hole model for magnetic phase diagram and pairing mechanism in copper-oxide superconductors. A. Robledo, C. Varea. *Rev. Mex. Fís.* **35**, 255 (1989) https://rmf.smf.mx/ojs/index.php/rmf/article/view/2050

II.xii.8 Magnetic vortex-antivortex pairing mechanism for doped La_2CuO_4. A. Robledo, C. Varea. *Physica C* **162-164**, 1517 (1989) https://doi.org/10.1016/0921-4534(89)90800-9

II.xii.9 Spin-hole model with magnetic vortex-antivortex pairing mechanism in copper-oxide superconductors. A. Robledo, C. Varea. *Physica C* **166**, 334 (1990) https://doi.org/10.1016/0921-4534(90)90414-A

II.xii.10 Magnetic anisotropy and superconductivity in a model for high-T_c copper oxides. A. Robledo. *Physica C* **220**, 271 (1994). https://doi.org/10.1016/0921-4534(94)90913-X

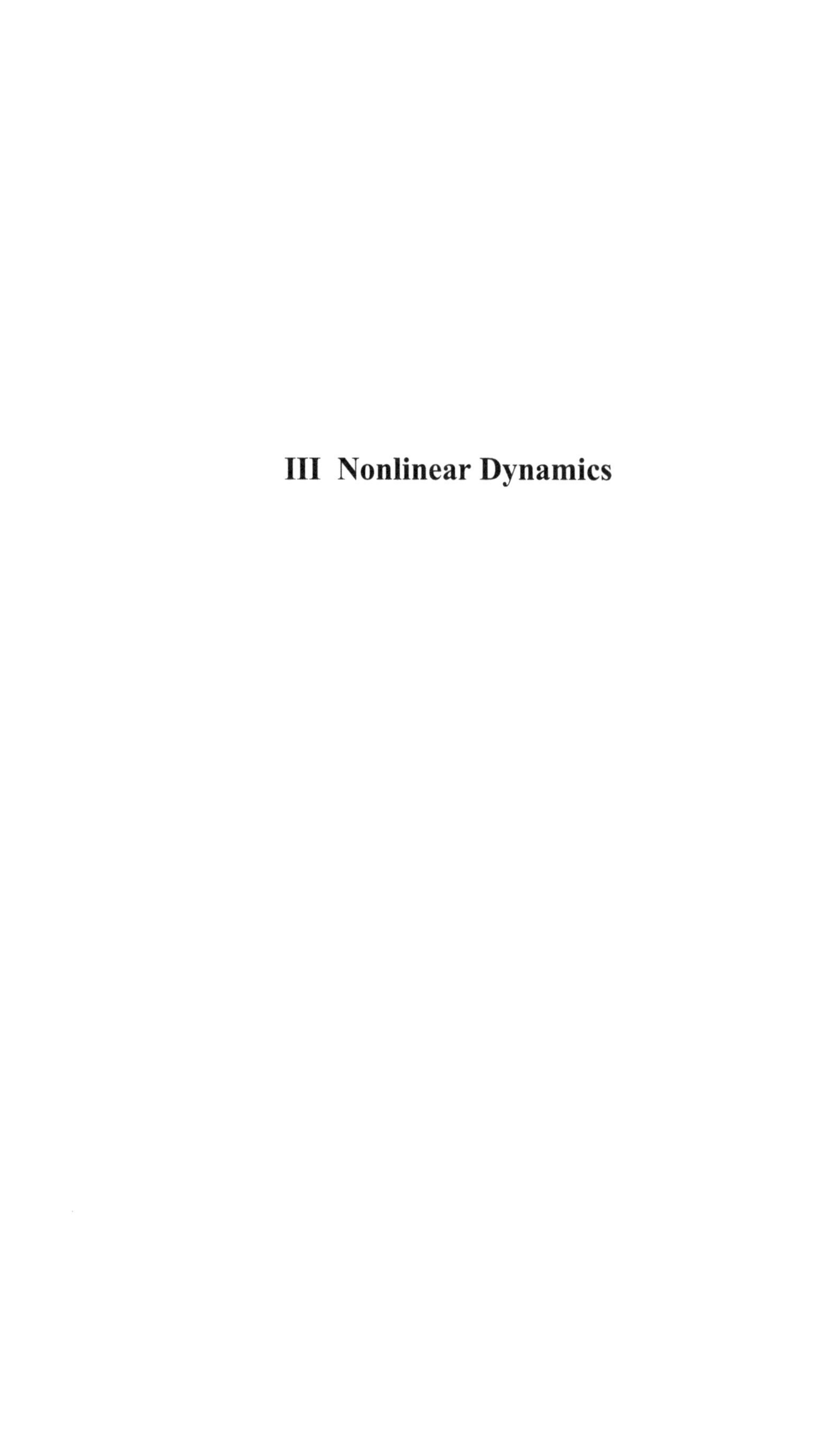

III Nonlinear Dynamics

Onset of chaos. A chance communication between Robert May and James Yorke in 1973 [47] led to the comprehension of the bifurcation diagram of the quadratic map, perhaps nowadays the better-known picture symbolizing nonlinear dynamics. This diagram illustrates the families of attractors or stable solutions for this difference equation and was first put together by linking the works of May [48] and Yorke and Tien-Yien-Li [49]. Not long after this, Feigengaum contributed [50, 51] to the full understanding of the universal scaling properties at the period-doubling accumulation point, the borderline between regular and irregular behaviors, the transition in and out of chaos.

Twenty-odd years ago we began, then increased and more recently consolidated, a linked set of research lines dedicated to the investigation of complex systems. This set rests on independent, original and rigorous developments on the transitions to chaos in dissipative nonlinear systems of low dimensionality, among which the quadratic map has been the most frequently used. Our investigations are based on two groups of properties we obtained and developed into effective research tools: the dynamics inside and the dynamics towards the attractors that represent the transitions to chaos. Based on this knowledge, studies on core problems of complex systems have been carried out. In the physics of condensed matter we have obtained results in problems of difficult treatment: glassy dynamics, localization and critical fluctuations. In complex systems we have achieved central results for evolutionary dynamics, ranked data distributions (Zipf and Benford laws), and on the rationalization and (needed) statistical-mechanical justification of the phenomenon of self-organization. Recently we demonstrated the equivalence of paradigms in this field, edge of chaos and criticality.

Our motivation to examine the transitions to chaos displayed by prototypical nonlinear iterated maps, such as the logistic and circle maps, originated from the goal to understand the possible existence of a limit of validity of the ordinary Boltzmann-Gibbs statistical mechanics and the conceivable appropriateness of the alternative entropy expression proposed by Tsallis [IV.ii.1]. The failure of both ergodic and mixing properties at these transitions, valid for their neighboring chaotic attractors pointed at a feasible 'numerical laboratory' in which to explore these issues. As a consequence of this all the studies described in this Part III contain results related to our quest of comprehending the so-called q-statistics.

With regards to complex condensed matter physics problems we mention the following: The joint use of density functional theory (for inhomogeneous systems) and the renormalization group fixed-point map (for the tangent bifurcation) to deliver a space-time description for the dominant fluctuation at a critical point. The bifurcation gap induced by addition of external noise, that removes the period-doubling and chaotic-band-splitting accumulation points, was shown to appear as a crossover phenomenon along time evolution at the onset of chaos. We found that before the crossover the dynamics is analogous to glassy dynamics in molecular systems. In addition a vital recursion relation for size growth of a basic wave scattering model was recognized as a nonlinear iteration map with a bifurcation diagram where tangent bifurcations separate periodic (insulating) and chaotic (conducting) attractors.

Amongst specific advances in nonlinear dynamics we make reference to: A central quantity in nonlinear dynamics, the sensitivity to initial conditions, was determined explicitly for the pitchfork and tangent bifurcations as well as for the period doubling and the golden ratio quasi-periodic transitions to chaos, all of which display anomalous properties while their Lyapunov exponent vanishes. The distributions for sums of successive positions of trajectories, as in random walks, for families of chaotic attractors of the quadratic map were found to conform to a renormalization group scheme such that its trivial fixed point matches the central limit theorem. From the start, the property that summarized our calculations for the dynamics towards the periodic attractors of the quadratic map resembled formally that of a partition function. Continued work supported this interpretation and finally a model for self-organization materialized.

With reference to contributions to the general understanding of complex systems we annotate: The initial consideration of the discrete-time version of the replicator equation of established models of evolutionary game theory lead straightway to novel bifurcation diagrams and valuable coupled-map models. The transformation of iterated map trajectories into (Horizontal Visibility) networks opened the possibility of reevaluating the link of the renormalization group technique with entropy optimization, and likewise the generalization of the Pesin theorem at the transitions to chaos. The size-rank distribution of all kinds of numerical data, including Zipf's law, were seen to be obtainable from nonlinear maps close to a tangent bifurcation, such that data samples are reproduced by their trajectories.

For reviews that contain descriptions of these developments see [52, 53, 54].

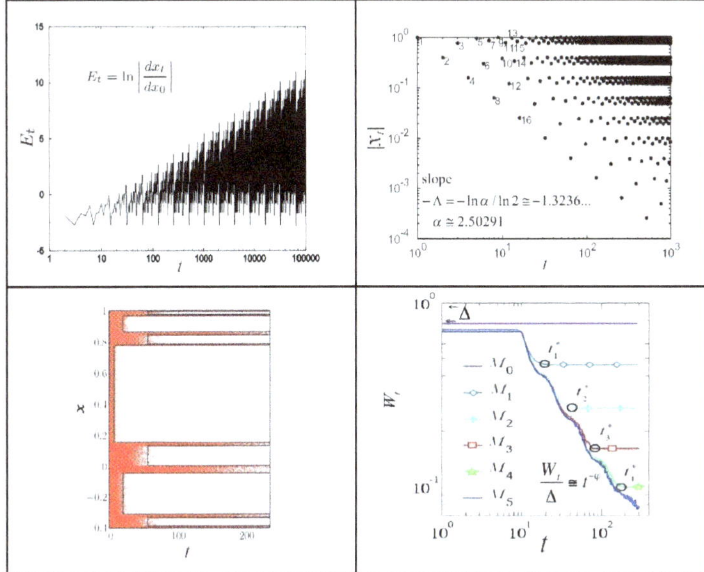

Figure 4: Dynamics at the transition to chaos. The top panels show anomalous properties at the transition to chaos in quadratic maps. In the left we see the expansion rate that implies vanishing Lyapunov exponent but rich behavior for the sensitivity to initial conditions. The right panel describes the interwoven power law structure of a trajectory inside the atractor. Section III.iv refers to studies related to these figure panels. The bottom panels show dynamical properties of an ensemble of trajectories running towards periodic attractors. In the left is seen how a family of gaps is formed sequentially by initially uniformly distributed trajectories. In the right is shown the power-law decay with logarithmic oscillations of phase space occupancy, where recapitulation is observed along the period-doubling family of attractors. Section III.vi refers to studies related to these figure panels.

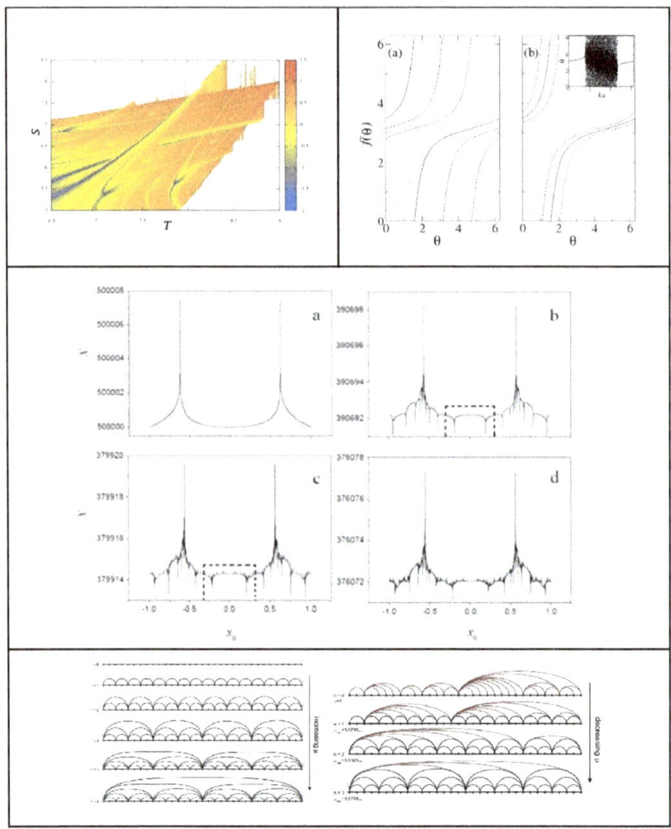

Figure 5: Modeling of complex systems via nonlinear dynamics. The left top panel represents in color the values for the Lyapunov exponent for the discrete time symmetric two-strategy cooperation games, including the prisoner's dilemma. Section III.vii refers to studies related to this figure panel. The right top panel illustrates maps close to a tangent bifurcation with properties equivalent to a model for localized to extended transport properties via scattering media. Section III.iii refers to studies related to these figure panels. The middle panel describes the development of a fractal function that represents sums of positions for ensembles of trajectories evolving under the action of period doubling attractors. Section III.v refers to studies related to this figure panel. The bottom panel illustrates the period doubling and chaotic band splitting attractor cascades as seen via (Horizontal Visibility) networks. Section III.viii refers to studies related to this figure panel.

III.i Critical fluctuations and the intermittency route out of chaos

We followed up studies [55, 56] for the spatial structure and temporal evolution of fluctuations at an ordinary continuous phase transition. These studies employed the Landau-Ginzburg-Wilson (LGW) effective free energy together with a deduced nonlinear iterated map near tangency. Detailed results were obtained for the dominant fluctuation, a large and relatively long-lived object obtained via the saddle-point approximation. More over, the time evolution of the dominant fluctuation was found to be of the intermittent type. The setting was ready for a possible statistical-mechanical application of the renormalization group fixed-point map at the tangent bifurcation, which would carry implications on its generalization known as nonextensive statistical mechanics. The employment of both the density functional theory (for inhomogeneous systems) and the renormalization group fixed-point map (for the tangent bifurcation) delivered a space-time description for the dominant fluctuation at a critical point [III.i.2–III.i.4]. Thus, we established an unforeseen connection between the area of critical phenomena in statistical mechanics and anomalous nonlinear dynamics at the transitions to chaos, a setting where generalized q-deformed entropy expressions appear naturally.

III.i.1 "Critical fluctuations, intermittent dynamics and Tsallis statistics". A. Robledo. *Physica A* **344**, 631-636 (2004). https://doi.org/10.1016/j.physa.2004.06.043

III.i.2 "Unorthodox properties of critical clusters". A. Robledo. *Mol. Phys.* **103**, 3025 (2005). https://doi.org/10.1080/00268970500185989

III.i.3 "q-statistical properties of large critical clusters". A. Robledo. *Int. J. Mod. Phys. B* **21**, 3947-3953 (2007). https://doi.org/10.1142/S0217979207045001

III.i.4 Dual characterization of critical fluctuations: Density functional theory & nonlinear dynamics close to a tangent bifurcation. M. Riquelme-Galvan, A. Robledo. EPJ-ST/172 Special Issue: Nonlinearity, Nonequilibrium and Complexity: Questions and Perspectives in Statistical Physics *Eur. Phys. J. Spec. Top.* **226** 433 (2017) https://doi.org/10.1140/epjst/e2016-60268-0

III.ii Glassy dynamics at the noise-perturbed onset of chaos

The bifurcation gap induced by the addition of external noise [3], that removes the finer features of the period-doubling and chaotic-band-splitting cascades together with their accumulation points, was shown to appear as a crossover phenomenon along time evolution at the onset of chaos [III.ii.1–III.ii.5]. It has been some time since we pointed out an unusual connection between two seemingly different situations [III.ii.1, III.ii.2], one of them the bifurcation gap and the other glass formation. The first belongs to nonlinear dynamics and the second to condensed matter physics. It was shown that they share main defining properties: the gradual disappearance of diffusion, the scaling law known as aging, anomalous two-step relaxation, etc. [III.ii.1–III.ii.5]. Noise amplitude in one setting represents temperature distance to a vitreous state in the other. The two problems indicate loss of ergodicity as noise amplitude or temperature distance vanishes. The quadratic map with additive noise can be seen to be a kind of discrete-time nonlinear Langevin equation. Nonetheless, in one case there is only functional composition while in the other there are molecular collisions. We have shown that the ideal glass concept exists and can be precisely represented by the attractor at the onset of chaos. This requires the knowledge that the noise-induced bifurcation gap is recapitulated at such, perturbed, onset of chaos. Before the crossover glassy dynamics is governed by q-statistics while after chaotic dynamics is reestablished ordinary statistics is recovered.

III.ii.1 "Universal glassy dynamics at noise-perturbed onset of chaos. A route to ergodicity breakdown". A. Robledo. *Phys. Lett. A* **328**, 467-472 (2004). https://doi.org/10.1016/j.physleta.2004.06.062

III.ii.2 "Aging at the edge of chaos: Glassy dynamics and nonextensive statistics". A. Robledo. *Physica A* **342**, 104-111 (2004). https://doi.org/10.1016/j.physa.2004.04.065

III.ii.3 "Intermittency at critical transitions and aging dynamics at edge of chaos". A. Robledo. *Pramana* **64**, 947-956 (2005). https://www.ias.ac.in/article/fulltext/pram/064/06/0947-0956

III.ii.4 "Glassy dynamics at the onset of chaos with additive noise". F. Baldovin, A. Robledo. *Fluctuation and Noise Letters* **5**, L313-L318 (2005). https://doi.org/10.1142/S0219477505002707

III.ii.5 "Parallels between the dynamics at the noise-perturbed onset of chaos in logistic maps and the dynamics of glass formation". F. Baldovin, A. Robledo. *Phys. Rev. E* **72**, 066213 (2005). https://www.doi.org/10.1103/PhysRevE.72.066213

III.iii Localization transition as a tangent bifurcation

A deep running analogy was uncovered between two apparently different physical problems, which allowed for the determination of elusive quantities and understanding of difficult issues. Intermittency and electronic transport were found to share common features and bridged the fields of research in nonlinear dynamics and condensed-matter physics. Specifically, the dynamics at the onset of chaos via intermittency appears associated with the critical conductance at the mobility edge of regular self-similar scatterer networks [III.iii.1]. More specifically, wave propagation through scattering media described by means of a double Cayley tree permits full solutions as the Bethe lattice, a form of mean-field, simplifies sufficiently the calculation of the scattering matrix. Other physical situations where the localization phenomenon occurs, light, sound or elastic media wave scattering, can be likewise modeled and described by nonlinear dynamics of low dimensionality, with the underlying implication of a drastic reduction of degrees of freedom [III.iii.1–III.iii.4]. This dynamics is represented by the Möbius transformation, the fixed points of which correspond to the localized states, and its ever-changing positions or phases to the extended states that display coherence due to vanishing Lyapunov exponent [III.iii.2]. Similar reductions of degrees of freedom leading to Möbius transformations have been observed in the synchronization of arrays of oscillators [57]. The conducting, coherent, weakly-chaotic regime exhibits q-statistics.

III.iii.1 "Equivalence between the mobility edge of electronic transport on disorderless networks and the onset of chaos via intermittency in deterministic maps". M. Martínez-Mares, A. Robledo. *Phys. Rev. E* **80**, 045201(R)-1-4 (2009). https://www.doi.org/10.1103/PhysRevE.80.045201

III.iii.2 "Möbius transformations and electronic transport properties of large networks". Yu Jiang, M. Martínez-Mares, E. Castaño, A. Robledo. *Phys. Rev. E* **85**, 057202-057202-4 (2012). https://doi.org/10.1103/PhysRevE.85.057202

III.iii.3 Typical length scales in conducting disorderless networks. M. Martínez-Mares, V. Dominguez-Rocha, A. Robledo. EPJ-ST/172 Special Issue: Nonlinearity, Nonequilibrium and Complexity: Questions and Perspectives in Statistical Physics *Eur. Phys. J. Spec. Top.* **226** 417 (2017) https://doi.org/10.1140/epjst/e2016-60129-x

III.iii.4 Invariant density of intermittent nonlinear maps descriptive of coherent quantum transport through disorderless lattices. V. Dominguez-Rocha, R.A. Mendez-Sanchez, M. Martínez-Mares, A. Robledo. *Physica D* **412**, (2020) 132623, https://doi.org/10.1016/j.physd.2020.132623

III.iv Generalization of the Pesin identity at the Feigenbaum attractor

A central quantity in nonlinear dynamics, the sensitivity to initial conditions, was determined explicitly for the first time for the pitchfork and tangent bifurcations as well as for the period doubling and the golden ratio quasi-periodic transitions to chaos [III.iv.1, III.iv.2, III.iv.5–III.iv.7, III.iv.9, III.iv.10], all of which display anomalous properties while their Lyapunov exponent vanishes. At the transitions to chaos in one-dimensional nonlinear maps the (only) Lyapunov exponent vanishes. A classical example is the period-doubling accumulation point attractor (the Feigenbaum attractor) that appears an infinite number of times in the bifurcation diagram of quadratic maps. There, the sensitivity to initial conditions behaves anomalously; it fluctuates endlessly in iteration time with increasing amplitude that grows sub exponentially. There is a remarkable property in the dynamics at the Feigenbaum attractor that leads to an identity between the q-generalized Lyapunov exponent and the rate of growth of the q-generalized entropy. This is the counterpart of the Pesin identity that states the equality of the (positive) ordinary Lyapunov exponent with the Sinai-Kolmogorov entropy for chaotic attractors. This property is that a distribution, say uniform, of initial conditions within a small interval, remains invariant for selected iteration times. The resulting q-generalized Pesin identity [III.iv.6] can be arguably considered to be the most important solid evidence regarding the so-called q-statistics [58].

III.iv.1 "Universal renormalization-group dynamics at the onset of chaos in logistic maps and non-extensive statistical mechanics". F. Baldovin, A. Robledo. *Phys. Rev. E* **66**, 045104-1-4 (R) (2002). https://www.doi.org/10.1103/PhysRevE.66.045104

III.iv.2 "Sensitivity to initial conditions at bifurcations in one-dimensional non-linear maps: rigorous non-extensive solutions". F. Baldovin, A. Robledo. *EPL* **60**, 518-524 (2002). https://doi.org/10.1209/epl/i2002-00249-7

III.iv.3 "The renormalization group and optimization of non-extensive entropy: criticality in non-linear one-dimensional maps". A. Robledo. *Physica A* **314**, 437-441 (2002).

III.iv.4 "Unifying laws in multi-disciplinary power-law phenomena: fixed-point universality and non-extensive entropy". A. Robledo. In: Non-extensive Entropy-Interdisciplinary Applications, Oxford University Press, C. Tsallis and M. Gell-Mann, editors (2004) pp. 63-78. https://doi.org/10.1093/oso/9780195159769.003.0008

III.iv.5 "Criticality in non-linear one-dimensional maps: RG universal map and non-extensive entropy". A. Robledo. *Physica D* **193**, 153-160 (2004). https://doi.org/10.1016/j.physd.2004.01.016

III.iv.6 "Nonextensive Pesin identity. Exact renormalization group analytical results for the dynamics at the edge of chaos of the logistic map". F. Baldovin, A. Robledo. *Phys. Rev. E* **69**, 045202(R) 1-4 (2004). https://www.doi.org/10.1103/PhysRevE.69.045202

III.iv.7 "Tsallis' q index and Mori's q phase transitions at edge of chaos". E. Mayoral, A. Robledo. *Phys. Rev. E* **72**, 026029 (2005). https://www.doi.org/10.1103/PhysRevE.72.026209

III.iv.8 "A recent appreciation of the singular dynamics at the edge of chaos". E. Mayoral, A. Robledo. in Verhulst 200 on Chaos, Understanding Complex Systems (Series), M. Ausloss and M. Dirickx, editores. (Springer, Berlin, 2006) pp. 339-354.

III.iv.9 "Dynamics at the quasiperiodic onset of chaos, Tsallis q-statistics and Mori's q-phase thermodynamics".
H. Hernández-Saldaña, A. Robledo. *Physica A* **370**, 286-300 (2006). https://doi.org/10.1016/j.physa.2006.03.018

III.iv.10 "Incidence of nonextensive thermodynamics in temporal scaling at Feigenbaum points and non-extensive thermodynamics". A. Robledo. *Physica A* **370**, 449-460 (2006). https://doi.org/10.1016/j.physa.2006.06.003

III.v   Renormalization group and central limit theorem for chaotic attractors

A claim [59, 60, 61, 62] about a novel kind of central limit stationary distribution for correlated variables to be displayed at the period-doubling onset of chaos attracted our attention and these led us to examine sums of positions of trajectories. This effort led us to clarify the issue by uncovering a remarkable renormalization group picture. The distributions for sums of successive positions of trajectories, as in random walks, for families of chaotic attractors of the quadratic map were found to conform to a renormalization group scheme such that its trivial fixed-point matches the central limit theorem [III.v.1–III.v.5]. In our work [III.v.1–III.v.5] we considered sums of positions from a single trajectory and also from an ensemble of them, the single trajectory at the transition to chaos leads to a multifractal-valued sum, that once re-scaled is similar to the trajectory itself. For ensembles of trajectories the structure of the fractal function, which is this sum at the onset of chaos as a function of the initial condition, is built in stages that recapitulate the additional increasingly finer features added along the period-doubling cascade. For chaotic band attractors the evolution of the distribution of sums of positions, of a single trajectory, or of an ensemble of them, inevitably takes the stationary

Gaussian form as the number of terms diverges. At the crossover from multifractal to gaussian behavior the distribution takes the form of a q-gaussian.

III.v.1 "Renormalization Group structure for sums of variables generated by incipiently chaotic maps". M.A. Fuentes, A. Robledo. *J. Stat. Mech. Theory Exp.* **2010**, P01001 (2010). https://doi.org/10.1088/1742-5468/2010/01/P01001

III.v.2 Stationary distributions of sums of marginally chaotic variables as renormalization group fixed points". M.A. Fuentes, A. Robledo. *J. Phys.: Conf. Ser.* **201**, 012002 (2010). https://doi.org/10.1088/1742-6596/201/1/012002

III.v.3 Sums of variables at the onset of chaos. M.A. Fuentes, A. Robledo. *Eur. Phys. J. B* **87**, 32 (2014) https://doi:10.1140/epjb/e2014-40882-1

III.v.4 Scaling of distributions of sums of positions for chaotic dynamics at band-splitting points. A. Diaz-Ruelas, M.A. Fuentes, A. Robledo. *EPL* **108**, 20008 (2014). https://doi.org/10.1209/0295-5075/108/20008

III.v.5 Sums of variables at the onset of chaos, replenished. A. Diaz-Ruelas, A. Robledo. *EPJ-ST Special Issue 160011: Temporal and Spatio-Temporal Dynamic Instabilities: Novel Computational and Experimental Approaches Eur. Phys. J. Spec. Top.* **225**, 2763 (2016). https://doi.org/10.1140/epjst/e2016-60011-y

III.vi Self-organization along the period-doubling cascade

The property that summarized our calculations for the dynamics towards the periodic attractors of the quadratic map resembled in form that of a partition function. Subsequent work gave support to this interpretation and finally a model for self-organization materialized from it [III.vi.1–III.vi.8]. We computed properties of ensembles of trajectories evolving towards (first periodic and then chaotic) attractors of the quadratic map. The expression descriptive of the space populated by trajectory positions at any given iteration time under the action of the attractor was identified as a partition function, equivalent to that of the construction by stages of a multifractal set. Ultimately, this statistical-mechanical likeness became a model for self-organization. The time evolution of the fraction of occupied space in logarithmic scales exhibits the telltale occurrence of discrete scale invariance, power-law decay dressed by logarithmic oscillations, and displays a 'recapitulation' property, i.e. evolution towards periodic or chaotic-band attractors repeats successively the evolution towards those attractors with smaller periods or number of bands. Only recently [III.vi.7, III.vi.8], the statistical-mechanical justification of the fraction of occupied space as a bona fide partition function was put together. The balance between numbers of configurations and Boltzmann-Gibbs statistical weights of the initial thermal system is strongly altered and ultimately eliminated by the sequential subdivision procedure that mirrors the actions of the attractor. However, the emerging set of subsystem configurations implies a different and novel entropy growth process that eventually upsets the original loss and has the capability of marginally [III.vi.8] locking the system into a self-organized state with characteristics of criticality, as in the so called self-organized criticality [63]. At the transition to chaos self-organization displays full scale-invariant properties similar to space and time scale invariance of critical states. There, the number of subsystem configurations and their generalized entropy, a q-entropy, is maximal [III.vi.8]. Attaining this state provides an explanation, within perhaps the simplest model system, for the hypothesis of self-organized criticality [63].

III.vi.1 "Some aspects of the dynamics towards supercycle attractors and their accumulation point, the Feigenbaum attractor". A. Robledo, L.G. Moyano. *AIP Con. Proc.* **965**, (2007) 114 https://doi.org/10.1063/1.2828722

III.vi.2 "q-deformed statistical-mechanical property in the dynamics of trajectories en route to the Feigenbaum attractor". A. Robledo, L.G. Moyano. *Phys. Rev. E* **77**, 032613-1-14 (2008). https://doi.org/10.1103/PhysRevE.77.036213

III.vi.3 "Labyrinthine pathways towards supercycle attractors in unimodal maps". L.G. Moyano, D. Silva, A. Robledo. *Cent. Eur. J. Phys.* **7**, 591-600 (2009). https://doi.org/10.2478/s11534-009-0065-1

III.vi.4 "Dynamics towards the Feigenbaum attractor". A. Robledo, L.G. Moyano. *Braz. J. Phys.* **39**, 364-370 (2009). https://doi.org/10.1590/S0103-97332009000400004

III.vi.5 "q-deformed statistical-mechanical structure in the dynamics of the Feigenbaum attractor". A. Robledo. *J. Phys. Conf. Ser.* **246**, 012025+6 9 (2010). https://doi.org/10.1088/1742-6596/246/1/012025

III.vi.6 "A dynamical model for hierarchy and modular organization: The trajectories en route to the attractor at the transition to chaos". A. Robledo. *J. Phys. Conf. Ser.* **394**, 012007-012007-9 (2012). https://doi.org/10.1088/1742-6596/394/1/012007

III.vi.7 Emergent statistical-mechanical structure in the dynamics along the period-doubling route to chaos. A. Diaz-Ruelas, A. Robledo. *EPL* **105**, 40004 (2014). https://doi.org/10.1209/0295-5075/105/40004

III.vi.8 Self-organization and a constrained thermal system analogue of the onset of chaos. A. Robledo. *EPL* **123**, 40003 (2018). https://doi.org/10.1209%2F0295-5075%2F123%2F40003

III.vii   Chaos in discrete-time game theory

Another case study we developed consisted of the inspection of the consequences of introducing discrete time to the replicator equation for a collection of well-known (social) games. We were headed straightforwardly into a nonlinear-dynamical extension of evolutionary game theory [III.vii.1]. We focused attention on the simplest option, the well-known social-dilemma (two-strategy, cooperation or defection) games represented by symmetric two-by-two payoff matrices. The appearance of chaotic dynamics in these games is ruled out by the Poincare-Bendixon theorem that establishes the requirement of at least three dimensions for the occurrence of chaos in a continuous time dynamical system. Therefore we chose to introduce discrete time into the replicator equation and convert it into a nonlinear iterated map with two control parameters [III.vii.1]. The results were immediate, the landscape of the three-dimensional bifurcation diagram uncovers a rich arrangement of periodic and chaotic attractors connected by recognizable but somewhat distorted period-doubling and chaotic band splitting cascades, windows of periodicity, etc. [III.vii.1]. Our nonlinear iterated map can be obtained after introduction of approximations to high-dimensional coupled-map models representative of evolutionary dynamics of ecosystems, like the Tangled Nature model [64]. In doing so we have obtained [III.vii.2, III.vii.3] with this approximation the possible connection between the macroscopic intermittent behaviors of the above-mentioned high-dimensional models with the known low-dimensional sources of intermittency, such as the tangent bifurcation. As a result q-properties at pitchfork and tangent bifurcations and at transitions to chaos have analogues in social games.

III.vii.1 Chaos and unpredictability in evolutionary dynamics in discrete time. D. Vilone, A. Robledo, A. Sánchez. *Phys. Rev. Lett.* **107**, 038101-1-038101-4 (2011). https://doi.org/10.1103/PhysRevLett.107.038101

III.vii.2 Tangent map intermittency as an approximate analysis of intermittency in a high dimensional fully stochastic dynamical system: The Tangled Nature model. A. Diaz-Ruelas, A. Robledo, H.J. Jensen, D. Piovani. *Chaos* **26**, 123105 (2016) https://doi.org/10.1063/1.4968207

III.vii.3 Relating high dimensional stochastic complex systems to low-dimensional intermittency?. A. Diaz-Ruelas, A. Robledo, H.J. Jensen, D. Piovani. EPJ-ST/172 Special Issue: Nonlinearity,
Nonequilibrium and Complexity: Questions and Perspectives in Statistical Physics *Eur. Phys. J. Spec. Top.* 226 341 (2017) https://doi.org/10.1140/epjst/e2016-60264-4

III.viii Complex network view of the routes to chaos

The transformation of iterated map trajectories into (Horizontal Visibility, HV) graphs opened the possibility of reevaluating the link of the renormalization group technique with entropy optimization, and likewise the generalization of the Pesin theorem at the transitions to chaos [III.viii.1–III.viii.7]. The examination of their resultant network structures, their degree distribution and their entropy expressions produced another significant result: the exposure of an HV network version of the q-generalized Pesin identity at the period-doubling [III.viii.3] and at the quasi-periodic [III.viii.6] transitions to chaos. In both cases, the 'refinement' quality of the HV algorithm, many time series into one graph, simplifies the multifractal set at the transition to chaos attractor into a fractal set. This advance required the network definitions for sensitivity to initial conditions and for Lyapunov exponent. The structure of the HV networks obtained from attractor trajectories lends itself in all cases to the consideration of a simple (contiguous-node-merging) renormalization-group (RG) transformation [III.viii.1, III.viii.2, III.viii.4, III.viii.5]. For the logistic map there are two trivial fixed-point HV graphs, those from period one (single chain) and from a single chaotic band (single chain dressed with random links), and one nontrivial fixed-point (scale-invariant) graph, that for the transition to chaos [III.viii.1, III.viii.2]. We have pointed out [I.iv.1] that there is a hidden entropy optimization procedure underlying the renormalization group technique, useful for extracting properties of systems with scale-invariant properties. Specifically, that the all-important trivial and nontrivial fixed points are extrema of a suitably defined entropy. The access to entropy is provided via the network degree distributions through the Shannon expression. The HV networks for period-doubling, quasi-periodicity and intermittency routes to chaos were determined.

III.viii.1 "Feigenbaum graphs: a complex network perspective of chaos". B. Luque, L. Lacasa, F.J. Ballesteros, A. Robledo. *PLoS ONE* **6**(9): e22411 (2011). https://doi.org/10.1371/journal.pone.0022411

III.viii.2 "Analytical properties of horizontal visibility graphs in the Feigenbaum scenario". B. Luque, L. Lacasa, F.J. Ballesteros, A. Robledo. *Chaos* **22**, 013109-013109-14 (2012). https://doi.org/10.1063/1.3676686

III.viii.3 "Feigenbaum graphs at the onset of chaos". B. Luque, L. Lacasa, A. Robledo. *Phys. Lett. A* **376**, 3625–3629 (2012). https://doi.org/10.1016/j.physleta.2012.10.050

III.viii.4 Quasiperiodic graphs: structural design, scaling and entropic properties. B. Luque, A. Núñez, F.J. Ballesteros, A. Robledo. *J. Nonlinear* **Sci.**, 23, 335-342 (2013). https://doi.org/10.1007/s00332-012-9153-2

III.viii.5 Horizontal visibility graphs generated by type-I intermittency. A. Núñez, B. Luque, L. Lacasa, J.P. Gómez, A. Robledo. *Phys. Rev. E* **87**, 052801-052801-9 (2013). https://www.doi.org/10.1103/PhysRevE.87.052801

III.viii.6 Quasiperiodic graphs at the onset of chaos. A. Núñez, B. Luque, M. Cordero, M. Gómez, A. Robledo. *Phys. Rev. E* **88**, 06918-1-06918-8 (2013). https://www.doi.org/10.1103/PhysRevE.88.062918

III.viii.7 Entropy and renormalization in chaotic visibility graphs, en Mathematical Foundations and Applications of Graph Entropy. B. Luque, F.J. Ballesteros, A. Robledo, L. Lacasa. *Wiley Online Library*, **6**, (2017) 1-39, https://doi.org/10.1002/9783527693245.ch1

III.ix   Universality classes of rank distributions revealed by nonlinear maps near tangency

The consideration of an existing stochastic approach for the reproduction of ranked data [65] led to a formal equivalence of a key mathematical expression with that for trajectories at the tangent bifurcation [III.ix.2]. This fact developed into a nonlinear dynamical approach for rank distributions that uncovered similarities with universality classes in critical phenomena [III.ix.1–III.ix.8]. The size-rank distribution of all kinds of numerical data, including Zipf law, were seen to be obtainable from nonlinear maps close to a tangent bifurcation, such that data samples are reproduced by their trajectories [III.ix.1–III.ix.7]. Remarkably, size-rank distributions with power law decay can be reproduced by trajectories of the renormalization group fixed-point map [III.ix.2]. And as it turns out also for distributions with exponential decay for which the point of tangency shifts to infinity. More generally, it was demonstrated that for all data source distributions the mentioned map can be constructed and the rank distributions determined [III.ix.7]. That is, the stochastic and the deterministic approaches are equivalent. This duality permits for an explicit and quantitative distinction between size-rank –sizes of cities– and frequency-rank –word frequencies– distributions, as the former appears as a trajectory while the latter is a sum of positions [III.ix.6]. The frequency-rank distribution turns out to be the functional inverse of the size-rank distribution [III.ix.6]. There are other surprising sets of properties related to this topic that can be obtained from the map at tangency. The reciprocals of the size-rank functions provide uniformly distributed probabilities for fixed rank that lead to extensive q-deformed entropies where system size is measured by sample size. Recently, we provided [III.ix.8] an extension to Number Theory as we obtain from the fixed-point map trajectories the numbers, or asymptotic approximations of them, for the Factorial, Natural, Prime and Fibonacci sets. Additional features of the parallelism with critical phenomena appear as borderline divergencies and logarithmic corrections for the Zipf law class. This in relation with the Prime numbers and exemplified by earthquake data. The formalism links all types of ranked distributions to a q-entropy.

III.ix.1 "Generalized thermodynamics underlying the laws of Zipf and Benford". C. Altamirano, A. Robledo. (Springer-Verlag, LNICST 5 2009) pp. 2232-2237, https://doi.org/10.1007/978-3-642-02469-6_100

III.ix.2 "Possible thermodynamic structure underlying the laws of Zipf and Benford". C. Altamirano, A. Robledo. *Eur. Phys. J. B* **81**, 345-351 (2011). https://doi.org/10.1140/epjb/e2011-10968-5

III.ix.3 "Laws of Zipf and Benford, intermittency and critical fluctuations". A. Robledo. *Chin. Sci. Bull.* **56**, 3643-3648 (2011). https://doi.org/10.1007/s11434-011-4827-y

III.ix.4 Incidence of q statistics in rank distributions. C. Yalcin, A. Robledo, M. Gell-Mann. *PNAS* **111**, (2014). https://doi.org/10.1073/pnas.1412093111

III.ix.5 Entropies for severely contracted configuration space. G.C. Yalcin, C. Velarde, A. Robledo. *Heliyon* **1**, (2015) e00045, https://doi.org/10.1016/j.heliyon.2015.e00045

III.ix.6 Rank distributions: Frequency vs. Magnitude. C. Velarde, A. Robledo. *PLoS ONE* 12(10): e0186015 (2017). https://doi.org/10.1371/journal.pone.0186015

III.ix.7 Dynamical analogues of rank distributions. C. Velarde, A. Robledo. PLoS ONE 14(2): e0211226 (2019). https://doi.org/10.1371/journal.pone.0211226

III.ix.8 Number theory, borderline dimension and extensive entropy in distributions of ranked data. C. Velarde, A. Robledo. PLos ONE 17(12): e0279448 (2022). https://doi.org/10.1371/journal.pone.0279448

III.x   Transition to chaos as critical point

The well-known properties of families of attractors of the quadratic map were revisited as seen through the densities (or measures) of ensembles of trajectories. These probability densities were determined via the Frobenius-Perron equation and trough them a novel statistical-mechanical picture was obtained [III.x.1]. Two families of attractors were considered, the supercycles along the period-doubling cascade and the Misiurewics points along the chaotic-band-splitting cascade, together with their common accumulation point at the transition to and out of chaos. From the densities the entropies associated with these attractors were determined and, most remarkably, when the collection of entropies for the two families of attractors is viewed along the values of control parameter the familiar pattern appears of a statistical-mechanical two-phase system separated by a continuous phase transition, an equation of state containing a critical point [III.x.1]. As we have already mentioned, the transitions to chaos and the fluctuations at a critical point display q-statistical properties.

III.x.1 Logistic map trajectory distributions: Renormalization-group, entropy, and criticality at the transition to chaos. A. Diaz-Ruelas, F. Baldovin, A. Robledo. *Chaos*, **31**, 033112 (2021). https://doi.org/10.1063/5.0040544

III.xi   Nonlinear dynamical view of Kleiber's law

The empirical law of Kleiber [66, 67], metabolism grows as mass to the 3/4th-power, has remained an open question for the life sciences since its formal finding in 1930. Why does this pattern hold for more than ten orders of magnitude that considers radically different organisms? Even though there have appeared plausible mechanistic models with quantitative agreements for both plants and animals [68, 69, 70], to understand this law we have taken the view of a universal, very general basis, as in the case of the power laws of critical phenomena. Recently [III.xi.1], we have reproduced quantitatively the available data for these two kingdoms in biology by means of a formalism that makes use of statistical mechanics and nonlinear dynamics. We consider RG fixed-point nonlinear iterative maps linked to an extensive q-entropy. We focus on a unique pair of universality classes that satisfy the 3/4th-power law, one of them corresponds to preferential attachment, rich gets richer, and the other to critical processes that suppress, overcome, the cost of motion. This conjugate pair of universality classes display, as the real data do, curvatures of opposite signs in logarithmic scales for small masses.

III.xi.1 A Nonlinear Dynamical View of Kleiber's Law on the Metabolism of Plants and Animals. L.J. Camacho-Vidales, A. Robledo. *Entropy* **26**(1), 32 (2023). https://doi.org/10.3390/e26010032

III.xii Bifurcation cascades within windows

The chaotic band attractors in the logistic map bifurcation diagram contain an infinite collection of intervals, called windows of periodicity. These windows contain sets of reproductions of the bifurcation diagram itself, and again, repeatedly, similar windows within each smaller replica, and so on, because the bifurcation diagram is a fractal object. The intervals display power-law spacing and widths that have been characterized a long time ago [71, 72]. There are other related structures in the bifurcation diagram, these are the so-called 'shadow' curves and their dual 'period' curves [73]. The shadow and period curves have mutual points of tangency and also self-intersections that envelop the bifurcation replicas inside the windows. The power laws displayed by the families of windows point at the existence of q-statistical-mechanical behaviors not yet characterized, as other sets of power laws occurring in the dynamics of quadratic and related maps have been, e.g. as in all the previous sections. These power laws correspond to the spacing of the windows across the bifurcation diagram, their widths, and other features within them, as are the replicas of the main cascades in the primary diagram. This complex structure is enveloped by the elaborate system of shadow and period curves. We are currently involved in the construction of models designed to describe complex systems where the most significant property is embedding, systems nested within systems.

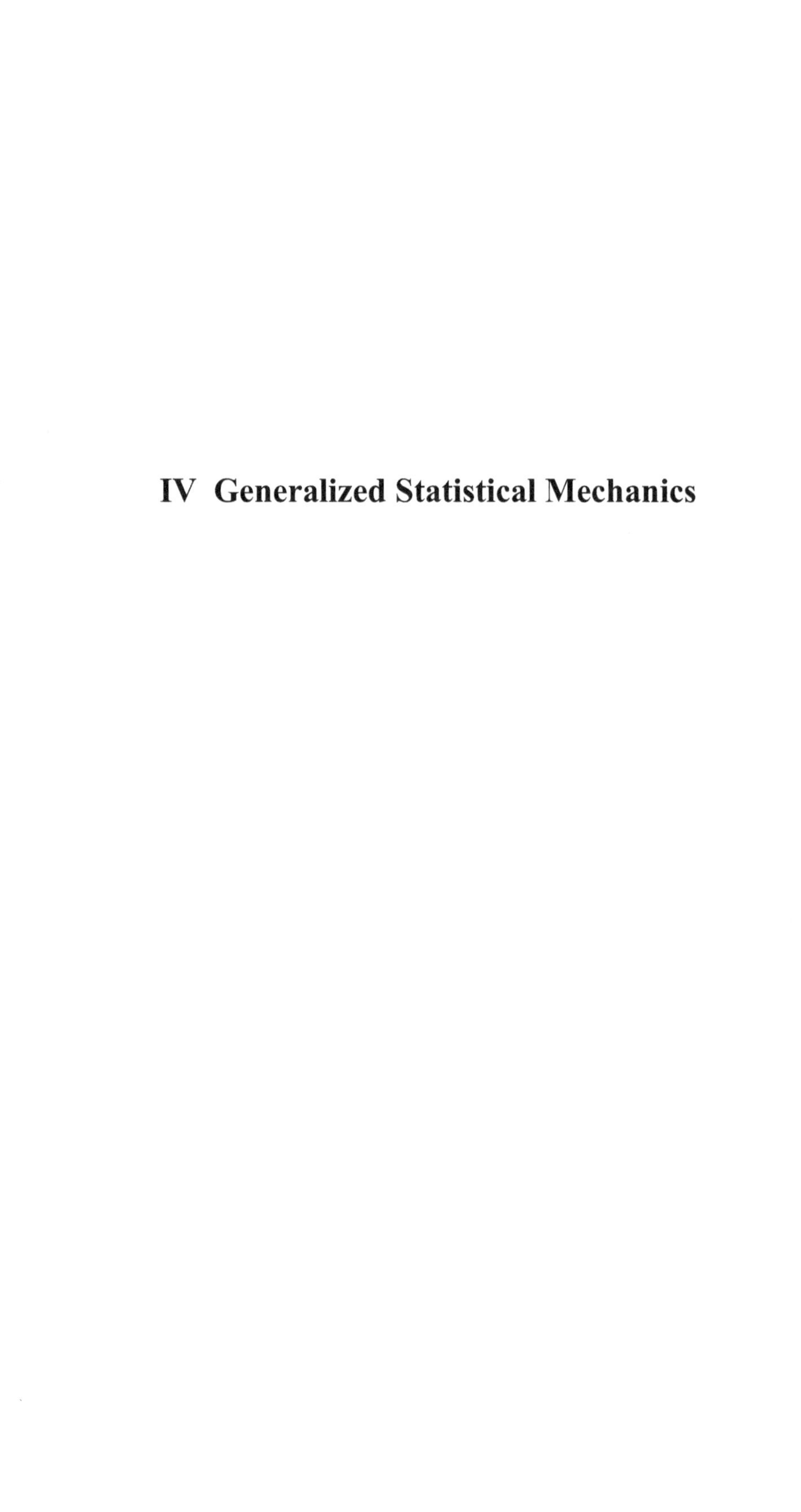

IV Generalized Statistical Mechanics

Limit of validity. As Constantino Tsallis himself describes [IV.i], it was during a workshop, held in 1985 at the Instituto de Física (Universidad Nacional Autónoma de México) in Mexico City, that he observed a professor pointing out the combination of symbols p to the power q typical of the description of multifractal sets. This made him consider a generalization of the familiar entropy expression in statistical mechanics and information theory. To his surprise, and others, the generalized expression preserves to a large extent the mathematical structure of the ordinary theory, or where there appear variants these are obtained without much difficulty. The initial publication of these developments [74] produced much interest that translated into many studies [58], but also disbelief and hesitation [75, 76]. Over the years this issue has developed into an evolving topic, under the names of nonextensive statistical mechanics or q-statistics [58] [III.iv.10], which carries with it an exploration of the fundamental basis of this branch of physics. It implies, if justified and pertinent, a limit of validity for the firmly established Boltzmann-Gibbs (BG) equilibrium statistical mechanics. Well-founded answers and thorough understanding have been awaited up to the present time while a large body of published studies has been assembled and it is still growing. We have developed an interest and have carried out research in this subject since the early stages and up to now. Our motivation has been two fold: The quest and substantiation of concrete instances in which the BG theory fails and the alternative theory is seen to be applicable. And relatedly, the understanding of the breakdown, if it is indeed the case, of the BG formalism, the underlying cause of the occurrence of a limit of validity of an otherwise unbeaten physical theory. Our early studies on this question involved our generalization of the Weierstrass random walks [I.iv.1, I.iv.5] that shows the connection between the RG technique and entropy optimization. A setting that we used to show the relationship of these walks with the anomalous dimension at criticality and where we found a role for the q-index of q-statistics [I.iv.1]. Then we made use of the same random walks to discuss the transit from ordinary to anomalous diffusion [I.iv.2] with reference also to transport in porous media [I.iv.3]. Our main effort, spanning two decades, to understand the implications of a generalized statistical mechanics has been focused on the field of nonlinear dynamics, specifically the transitions to chaos in dissipative systems. This choice was and is motivated by the fact that the chaotic attractors present in these systems possess the fundamental properties of BG statistical mechanics: ergodicity and mixing, and that these two essentials breakdown at the borderline situation, the transition attractors from chaotic to regular behavior. The additional feature of relative simplicity of (precise, exact) access to their properties makes low-dimensional mathematical models desirable 'numerical laboratories' for this purpose. The entire contents of Part III of this research report, its twelve sections, have produced concrete advances related to the q-statistics issue, at the same time these studies have revealed the areas in condensed matter physics and in complex systems where the generalized statistical-mechanical formalism is applicable. See the review articles in Refs. [52, 53, 54]. For an earlier account of the initial stages of q-statistics and of our motivations to participate in the development of this subject see Ref. [IV.i] But it was only very recently [IV.ii] that several apparently separate pieces of research came together to create an important advance in the understanding of how the limit of validity of the BG theory arises and in what way it is replaced by, precisely, q-statistics. This fusion or amalgamation requires first a change in perspective when considering the familiar phase space variable x in one-dimensional nonlinear iterated maps $f(x)$. Instead of seeing such maps as systems composed of only one degree of freedom, they can be seen to represent systems with many, infinite, degrees of freedom, when described thermodynamically through a macroscopically observable variable like the energy or the magnetization, the map variable x. In other words, from this viewpoint the map includes already a (generalized) thermodynamic quantity, the result of the familiar average of (microscopic) configurations in statistical mechanics that results in (macroscopic) observables. Then the next step is to consider not only stationary (equilibrium) states but also processes that lead to these final states. We resort to the rate equations (that we have referred to in Section II.iii) for the description of the kinetics of phase change in condensed matter physics, such as the dissipative Landau-Ginzburg equation [25, 30] [II.iii.1] discussed here in Part II. We focus on three quantities: the differential equation variable, the driving force of the equation, and its Lyapunov function [77][II.iii.1]. After this we show that the discrete-time version of the Landau-Ginzburg equation is a dissipative nonlinear iterated map $f(x)$ and that the general choice of the Renormalization Group (RG) fixed-point maps for the three known routes to chaos capture a wide range of possible applications (including those described in the twelve sections of Part III). In addition to all this we demonstrate that the RG fixed point maps and all of their trajectories have analytical closed q-exponential expressions. Technically, the results of this approach are: i) The q-exponential function as the fabric of the RG fixed-point maps for the three routes to chaos. ii) Its inverse function the q-logarithm, identified as the Tsallis entropy expression, that, significantly, turns out to be the Lyapunov function of the discrete time Landau equation. As for a brief explanation of the limit of validity of ordinary statistical mechanics and the subsequent incidence of Tsallis statistics: These are a consequence of the effect of the RG fixed-point maps attractors in reducing drastically the initial phase space, or in terms of the phenomena modeled by them, in a severe hindrance to access initially available configurations. See Ref. [IV.ii] for a detailed analysis and explanation.

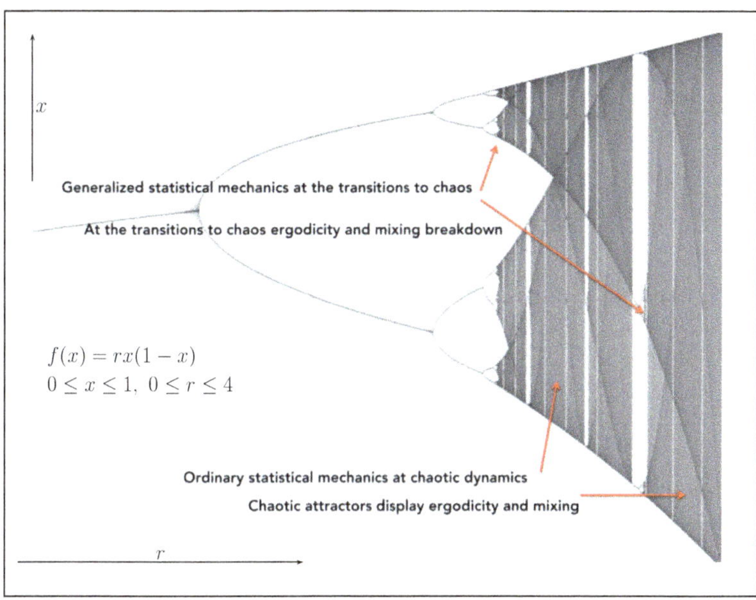

Figure 6: The quadratic map bifurcation diagram. This well-known diagram represents our numerical laboratory dedicated to the study of the limit of validity of ordinary equilibrium statistical mechanics and to the examination of the reliability of its generalization, q-statistics, at the transitions to chaos attractors. See Sections IV.i.1 and IV.ii.1.

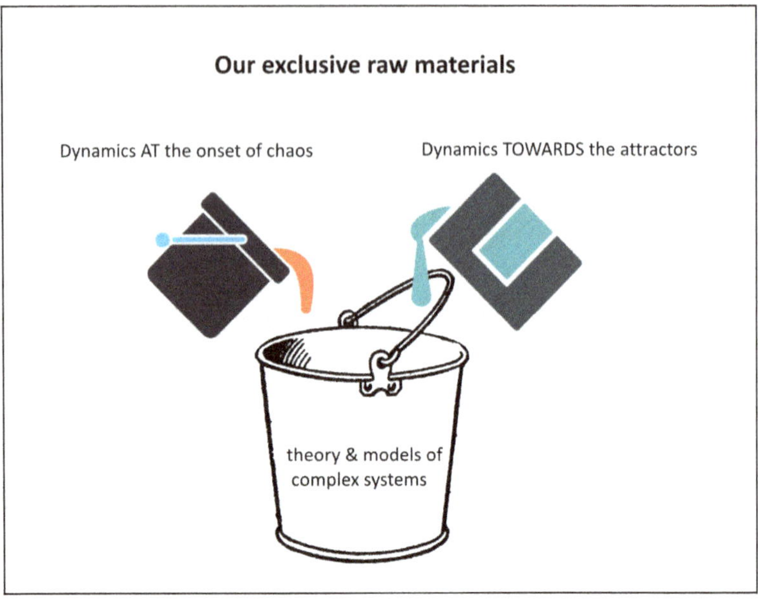

Figure 7: Cartoon that shows the ingredients we use to model complex systems. The models address current questions about complexity described in Part III. We employed the dynamics within the attractors at the transitions to chaos, and also the dynamics towards these attractors, to exhibit the applicability of q-statistics in our complex system models. See Sections IV.i.1 and IV.ii.1.

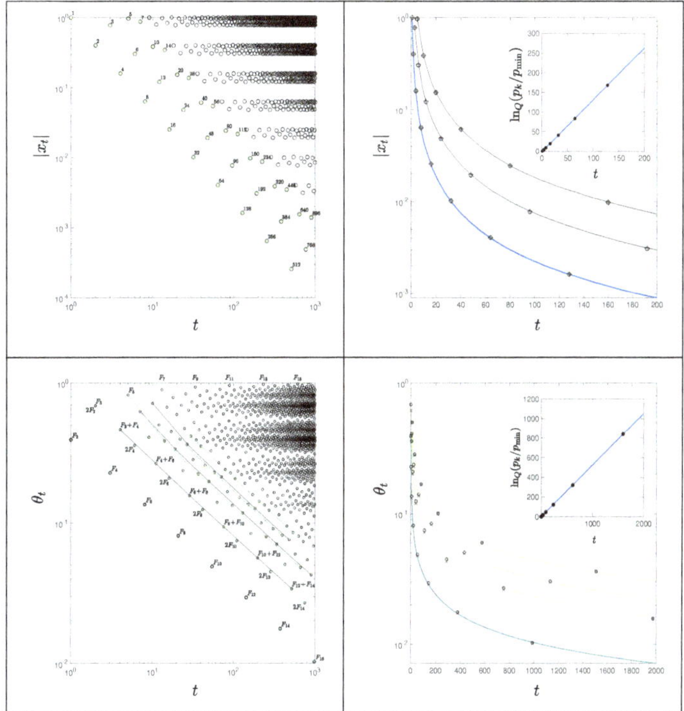

Figure 8: Principal trajectories within attractors at transitions to chaos. Top left panel via period doubling and bottom left panel via quasi-periodicity for the golden ratio irrational number. Both in logarithmic scales to highlight the power-law sets of "diagonal" positions. See Section IV.ii.1. Top right and bottom right panels reproduce the diagonals via the Renormalization Group (RG) fixed-point map for the transition to chaos via intermittency. The q-exponential closed-form analytical expression for this fixed-point map and its trajectories lead to two important results: i) The finding of previously unknown analytical expressions for the RG fixed-point maps for two of the three routes to chaos in one-variable maps. ii) The discovery of the formal link of the dynamics at those transitions to chaos with q-statistics. See Section IV.ii.1 and reference cited there.

IV.i A new statistical mechanics

If fully chaotic dynamics lays the foundation for statistical mechanics, the question arises: What happens when dynamics is on the edge of becoming chaotic? What are the limits of validity of traditional statistical mechanics? Is there any generalization of the existing theory?

Figure 9: Photograph of Alberto Robledo taken around the time of elaboration of the publication cited in Section IV.i.1.

The genesis of a new statistical physics at the edge of chaos*

In the sphere of physics, the standard method for describing the collective properties of a large numbers of particles —known as statistical mechanics— has been enormously successful for more than a century. At the same time, it has suffered from restrictions in its scope, since it only applies to a limited set of circumstances: the so-called equilibrium states. Developed in the nineteenth century, statistical mechanics enabled physicists to overcome a formidable problem. Common substances, such as water or iron, composed of myriads of atoms, make it impossible to calculate in perfect detail the

---

*Original in Spanish [IV.i.1]. Translated by Luis Jovanny Camacho-Vidales. During the two decades since the original version of this account was written, a large number of studies related to the particular goal expressed below have been published [52, 53, 54]. Very recently, a conclusive answer to our general quest was uncovered, established, and published [IV.ii.1].

temporal evolution of the particles multiple arrangements or configurations. However, access to an understanding of their behavior becomes accessible by giving up a detailed description and instead proceeding to work solely with probabilities. The knowledge of the probabilities associated with the particles forming all their possible configurations, and for most circumstances, only averages obtained from them, are sufficient to obtain all the equilibrium properties of the system.

Subsequently, research advances in the theory of dynamical systems [3] in the last 30 years or so have yielded important insights into the hypotheses underlying statistical mechanics. The non-linear nature of the equations of motion of the particles or molecules of physical systems has allowed us to conclude that the profoundly irregular nature of their trajectories, technically named chaotic, precisely produces the conditions that validate statistical mechanics. For example, quantities of interest, such as molecular velocities, obey normal statistics, and their distributions show the traditional Gaussian bell shape. The chaotic quality of particle trajectories also makes it easier to understand why systems, when left to evolve on their own, tend to become disorganized, displaying irreversibility in time and memory loss of its initial conditions. In technical language, they acquire the properties of ergodicity and mixing in their configuration space.

If fully chaotic dynamics lays the foundation for statistical mechanics, the question arises: What happens when dynamics is only incipiently chaotic? What are the limits of validity of traditional statistical mechanics? Is there any generalization of the existing theory? What new problems can be addressed under these extreme circumstances? It is in this context that the results of our investigations play a part. These studies refer to nonlinear dynamical systems, simple in their definition but complex in their behavior, which shows episodes occurring between regular and chaotic movement, the so-called onset of chaos. Its properties are of special interest as they show parallels with systems of many degrees of freedom. In particular, the three famous routes to chaos (via intermittency, via duplication of periods and via quasi periodicity), which are observed in the standard logistic and circle "mappings" or "maps". These routes display anomalous dynamics with universal characteristics that have been understood only recently. Our contribution in this field has been the rigorous elucidation of the properties of incipient or weak chaos in these systems, and their possible implications for the foundations of statistical mechanics.

The results of our studies suggest that there is indeed a form of statistical mechanics that generalizes the traditional one and that is applied when the system does not have access to all of its microscopic configurations or suffers impediments to fully "explore" these configurations over time. In these extreme cases, exemplified by the hindered dynamics associated with glass formation, the time evolution shows "infinite memory" and an absence of the basic properties of ergodicity and mixing. Its mathematical description differs from the usual exponential function (the Boltzmann factor or the Gaussian distribution) and instead accepts the so-called "power laws", more specifically a deformation of the exponential, the "q-exponential" function that resembles for large arguments a power law [80] The subset of accessible configurations forms in these cases a restricted space with a singular geometry that corresponds to objects called "fractals".

An entropic revolution

As told by Constantino Tsallis, the famous Brazilian physicist, during a scientific event in which he participated in the Institute of Physics at our National University in 1985, one of those days he observed a discussion between a student and a professor during which the combination of symbols p to the power q, typical in the theory of geometric objects called multifractals, appeared repeatedly on the blackboard. This led him to consider an expression in terms of this combination, with p probability and q an index with meaning still pending, as the generalization to the customary expression for entropy in statistical mechanics and information theory. To his surprise, and that of many, with this novel expression, the same mathematical structure of the usual theory, free energies, Maxwell relations, etc., etc., is recuperated to a great extent, and without great difficulties. When the parameter q takes the unit value $q = 1$, the canonical theory is recovered. Tsallis published his first advances in this direction in 1988 [74], and an intriguing option arose to explore aspects of statistical physics from a new perspective.

In the decade following the publication of this initial work, interest began to be generated both in analyzing and extending the structure of the theory, and in exploring which physical systems, and under what circumstances, would the application of this generalization of statistical mechanics be appropriate. Regarding the first effort, an important point has been the examination of the principle of maximum entropy. We recall that the famous Central Limit Theorem states that the probability for the sum of a number N of independent random variable outcomes converges to a Gaussian distribution as N increases indefinitely. Notably, among all the probability distribution functions that share the same width (variance), the entropy given by the standard expression acquires its maximum value for the case of the Gaussian distribution. Similarly, a central element of thermodynamics and statistical mechanics is the principle of maximum entropy, which states that for an isolated system, therefore with a constant number of molecules and constant energy, assigning the same probability to all configurations maximizes entropy. However, the analysis and deduction of the equivalent properties for the new theory were not completely successful, as certain difficulties arose that have only now been better understood. One of them, for example, concerns the maximization of the new entropy under the assumption of a fixed value of some quantity. In this case, another fixed quantity other than energy has to be assumed, a "generalized energy" which lacks a transparent physical interpretation. This modification is due in order to have a unique value for the q index for both the probability distribution and the entropy. This situation arises due to the way in which the average of the energy is considered in the maximization of this new entropy. The ordinary energy average leads to a "duality" of indices, q and $2 - q$.

At this time, the understanding of the "duality" between the values of the parameter q points the way to the resolution of this apparent puzzle [III.iv.6]. As for the possible applications of the new formalism, a multitude of propositions in physics and in many other disciplines arose, almost all of them associated with the ubiquity of the so-called power laws. These cover topics as diverse as complex fluids, the distribution of galaxies, encephalography, the swings of stock market shares in economics, the relationship between the frequency and intensity of earthquakes, etc. But we do not have enough space here to describe its virtues and sins [81].

The next stage in the history of this singular scientific process was the production of a large number of studies by a growing number of physicists based in different countries [82]. This progressive development gave visibility to the subject on international scale. Three large areas of research began to stand out in systems where access to their configurations is possibly conditioned or restricted: states far from equilibrium, such as turbulent flow. In this field, the work that has been carried out is of phenomenological nature and the quantitative agreement achieved with the experiments is difficult to rationalize in the absence of better evidence. A second area of research was forged in the study of models of fluids and magnets or gravitational systems whose elemental components interact with long-range forces. Here, the system is clearly defined at the microscopic level so that the statistical mechanical problem is well stipulated. The evidence of applicability of the new statistics is suggestive, but the way to reach firm conclusions is limited by the need of numerical simulations without feasibility of applying an analytical methodology. The third area of research has focused on the study of nonlinear dynamical systems at the onset of chaos, the borderline between ordered and irregular behavior. In this topic, the models are simple in their definition, but capable of showing a great variety and complexity of behaviors. Initially, numerical studies were carried out that suggested the relevance of the new theory in these problems. These models are nonlinear iterated maps amenable to rigorous analysis and they offer for this reason an ideal opportunity to judge the validity of predictions coming from any formalism. Included in this evolution of the subject criticisms emerged on plain aspects of this new statistical mechanics. For example, it was pointed out that apparent success in many applications rests on the ability to adjust parameters to achieve quantitative agreement. The optimization of entropy generates a distribution whose functional form is the mentioned q-exponential. This function approximates a power law for large argument. Then, the reproduction of experimental data corresponding to the abundant phenomena described by power laws is obtained by fixing the value of the enigmatic index q without generating any genuine understanding of its meaning. Another type of criticism refers to the internal consistency of the theory, for example: How does it perform in the case of systems at equilibrium? (Assuming that it is applicable to such circumstances). How does the concept of temperature, etc., operate under these new circumstances? Some other criticisms refer to the historical precedence of the generalized expression for entropy. It has been pointed out that it was first coined and used by Havrda and Charvat many years in advance [83]. By coincidence, I was able to learn from an Indian colleague, at Jawaharlal Nehru University in New Delhi, India, Professor Karmeshu, that he made Tsallis aware, several years after 1988, of these authors' work published in Kybernetica. Tsallis carried out his generalization independently and gave to it a very different direction.

I first met Constantino Tsallis many years ago, in 1982 or 1983, at a physics meeting in San Jose, Costa Rica, when he and I were interested in other topics in statistical physics. Later I invited him to an event in Oaxaca in 1995 when these new ideas in statistical physics were already in full development. It was a little later, around 1999 that I decided to undertake studies on my own in order to intervene in the development of this field of research with a certain purpose in mind. It was clear to me that a deeper and more rigorous understanding of this issue was required, that the statistical physics community required more substantial analysis and results to form a stronger opinion. The result of this search could point out very different conclusions. One possibility was that the whole scheme remained a mathematical curiosity, aesthetically appealing but with no real justification in nature. Or, at the other extreme, the clarification and understanding of the reasons for its applicability to its true field of action. On the one hand, it was necessary to unravel the meaning of the new parameter q and show how it could be determined by the fundamental variables in each problem where it found validity. Some questions were: Does the parameter q take only fixed values or a continuum of values? How do you determine these special values? What is the relationship between the parameter q and the ability of the system to access all its configurations? Is q different from the unit value when the basics of the standard theory fail? It seemed to me that the most effective path would be to rigorously elucidate a single example, a specific system or model, where the ingredients of loss of ergodicity and mixing were present. On the one hand, the problem should be sufficiently known, central to statistical physics or to the theory of dynamical systems and not an exotic problem outside their confines. On the other hand, the problem should be simple enough, though not trivial, so that it could be solved without approximation, analytically, and open to examination and scrutiny at leisure. Of a nature to be easily brought up in detailed discussions with professional colleagues with both attitudes, for and against. This would be a first step to establish the appropriateness or irrelevance of the theory. I selected the problem of dynamics at the onset of chaos in simple nonlinear iterative maps, which for specialists were considered by then perhaps clichéd or worn-out.

The physics involved in circumstances of loss of ergodicity and mixing have the potential to be truly interesting. It is known that power laws and the consequential property of "scale invariance" arise in incipiently chaotic systems. Under these circumstances, small perturbations to a system may not be enough to alter the arrangements of its microscopic constituents, such as the molecules of a fluid or crystal. Consequently, the system is unable to explore all its possible configurations for long enough periods of time. In contrast, in fully chaotic systems even the slightest perturbations keep the system cruising towards new configurations, allowing the exploration all its states as required by Boltzmann statistics. This efficient way of exploring the configuration space obeys the same law as the well-known Brownian movement, or ordinary diffusion, as observed when a drop of ink spreads out in crystalline water until it clouds the entire container. There are two kinds of systems with the capability to represent this situation: Anomalous random walks and the onset of chaos in iterative mappings. The former deviate from Boltzmann statistics since their probability distributions differ markedly from the Gaussian law associated with Brownian motion. In the latter case, the configuration space is drastically restricted to a fractal set, or more precisely a multifractal set (which we define later). We conducted investigations for both cases. Regarding the first, we only indicate the bibliographical references here [I.iv.5]. Next, we briefly refer to our current progress and findings for the second case.

The edge of chaos

The prominence in contemporary physics of "critical attractors" present at the edge or onset of chaos in the so-called logistic and circle mappings and many other similar mappings comes from their remarkable universal properties. These are comparable with those that govern the critical phenomena in systems with many degrees of freedom such as phase transitions associated with fluids, magnets, superconductors, etc. In critical attractors, indicators of chaos disappear, such as the rapid separation of initially very close orbits (or trajectories). Instead, at the onset of chaos, trajectories recede and recapitulate approximately, incessantly, maintaining memory of their previous positions and tracing intricate patterns that manifest certain regularities during their evolution. The dynamics under these circumstances acquires unexpected symmetries consistent with those displayed by fractal objects or, more precisely, multifractals. A fractal object is a geometric object whose basic structure is repeated at different scales. In many cases, fractals can be generated by a recursive or iterative process capable of producing self-similar structures regardless of the specific scale. Fractals are geometric structures that combine irregularity and structure. Its dimension is in general a non-integer greater than its topological dimension. Unlike geometrical fractal models, whose definition is given in mathematical terms, fractals in nature, such as clouds, mountains, and blood vessels, have lower and upper limits on their detail. A multifractal is a multitude of fractals in a single object, or equivalently, a probability distribution defined over a fractal set [84]. The renowned onset of chaos is already a classic example of how the use of simple nonlinear models, say the logistic mapping, as a starting point, often leads to important advances in the theory of dynamical systems. The discovery of their universal properties triggered a burst of activity more than two decades ago, and by now the ways in which ordered periodic behavior becomes chaotic or irregular are widely understood and experimentally verified. However, the nature of the dynamics in this border state, the transition to chaos, was only partially understood over time. On the contrary, the geometric aspects of its attractor, such as those of the so-called Feigenbaum attractor that appears countless times in the logistic map, have been well understood and these days the tools forged to carry out this undertaking are the basis of the study of these interesting objects. However, as we mentioned, the dynamical aspects have remained open to better understanding. It is precisely in this niche where a situation incapable of being described by traditional statistical mechanics remained hidden. Our investigations had the role of breaking down this phenomenon in detail and of rigorously verifying the validity of the ideas on which the proposal of a new statistical physics rests. Findings on dynamics at the transitions to chaos also have universal properties and we will only touch on them briefly. The interested reader can find out more detail in the references provided below [III.iv.1 to III.iv.9]. We recall that trajectories in deterministic chaos do not tend asymptotically to a fixed value at large time, nor to a periodic orbit, or to a quasi-periodic orbit. Chaotic trajectories that start close spread out exponentially. It is called deterministic chaos because the system does not contain random terms or parameters dependent on a probability distribution, and therefore the irregular behavior arises from nonlinearity. An attractor is a set of positions where all trajectories converge. Fixed points, periodic limit orbits, and chaotic limit trajectories are examples of attractors. These show general ingredients: Any trajectory that is in an attractor will be in it for all iteration times, additionally, they attract sets of trajectories with different starting positions. The set of these positions is called its basin of attraction. The simplest type of attractor is the fixed-point attractor. The system that has a fixed-point attractor will tend to stabilize at a single point. A common example is the pendulum, which due to friction with the air tends to the point where the angle is zero with respect to the vertical. The limit cycle (or periodic) attractor is the second simplest type of attractor. This type of attractor tends to stay in the same period forever. As an example, we can take a powered pendulum to counteract the friction force so it would swing from side to side. In these two cases the trajectories converge exponentially and this defines a negative value for the "Lyapunov coefficient". The trajectories associated with the chaotic attractor diverge exponentially so that its Lyapunov exponent is positive. This number is an indicator of chaos. For a critical attractor, such as at the transitions to chaos, the Lyapunov exponent is zero. The new theory presupposes a generalization of this quantity.

The research enterprise we conducted was first focused on digging deeper into the anomalous dynamics at the transitions to chaos via the above-mentioned three routes or pathways. To this purpose we considered the well-known logistic and circle maps. We discovered that the temporal evolution at the transition to chaos resembles the structure of the entire route to chaos, that is, right at the onset of chaos, the sequence of periodic attractors is reproduced asymptotically, ordered in larger and larger periods until they diverge to infinity. In the past, this sequence of periodic attractors was studied, one by one, by varying the control parameter of the map. In our case, we tuned the said control parameter precisely at the onset of chaos and unraveled the structure of its orbits. This structure evolves over time generating a progressive image of the periodic orbits that occur before the onset of chaos that appear sequentially through bifurcations from its smallest to its largest periods. In terms of an allegory, it could be said that this cascade of events is what has been said in the literature about the moment of the death of a human being. That at that moment he/she remembers his/her whole life from childhood to that moment. We deduced our results with the assistance of the technique called "renormalization group", which is designed to analyze problems with the property of self-similarity or "scale invariance". This technique has been very successful in the study of critical states in condensed matter and nonlinear dynamical systems.

Once the secret of the temporal organization of trajectories at the onset of chaos was known, we carried out the next step. This consisted in the determination of the function that measures the separation of trajectories in time, the sensitivity to initial conditions, which could be compared with the form suggested by the new statistical physics. We were able to carry out this task exactly and we found, on the one hand, that, yes, indeed, the new statistics were in line with our results. Although we found a much more complex reality than originally anticipated. Instead of a sensitivity to initial conditions described by a single function with power law property (or equivalently "q-exponential"), we found an infinite number of them, intertwined, forming a family that shares the same special value of the parameter q, but each one with a different generalized Lyapunov coefficient, so that the evolution through them forms a specific pattern. Furthermore, the entire dynamics inside the attractor is constituted by an infinite set of these families, each one with its q value. Regarding the physical origin of these special values of the parameter q, we established that they arise due to the presence of dynamical phase transitions. These transitions correspond to trajectories that associate different regions of the multifractal attractor with the important property that the values of q are universal constants for the route to chaos under consideration. These families of transitions are ordered in a hierarchical way, one of them dominates, the one that links the most condensed region with the most rarefied region of the multifractal. Our results cover all three routes to chaos. The above description applies to routes via doubling of periods and via quasi-periodicity. For the path via intermittency, the situation is considerably simpler than the one described here since in this case the critical attractor is not a multifractal. For further details or explanations, consult the references provided.

Connections with problems in condensed matter physics

In addition to the developments described, we have extended our investigations to the study of phenomena of importance in condensed matter physics by establishing connections between the dynamics at the transitions to chaos with several phenomena. These correspond to phase transitions, such as the critical point of the condensation of a gas into a liquid, or the transformation of an insulating material into a conductor, or to extreme situations such as the formation of the glass state. The connections we studied suggest that the presence of a new statistical physics in states of incipient chaos in non-linear dynamical systems is also manifested under certain circumstances in the more complex systems of condensed matter physics. We have developed three lines of work, one for each route to chaos. In the case of the route via intermittency, we found that the temporal evolution of clusters or fluctuations in order parameter (magnetization in a magnet or difference in densities between gas and liquid in a fluid) in critical states precisely obeys the anomalous dynamics described by the trajectories of the map [III.i.1, III.i.2]. See also [III.i.3, III.i.4] In the case of the route via period doubling, we made a remarkable finding consisting of a deep analogy between the slow dynamics shown by supercooled liquids in the vicinity of glass formation and the dynamics of the critical attractor when it is externally disturbed by "noise", that is, a stochastic agent [III.ii.1 to III.ii.5]. In the case of the route via quasi-periodicity, we took advantage of the equivalence between the transition from localization to transport in incommensurable materials and the dynamics of the critical attractor associated with this kind of incipient chaos. The localization transition describes the transformation from an insulating material to a conductor of electric charges when it undergoes structural transformations that move it away from the crystalline state.* These examples constitute a base from which the applicability of the new statistical physics can be founded. Given the space and purpose of this story, we briefly comment below only the case of critical fluctuations.

The condensation of a gas into a liquid is a natural phenomenon that is very familiar to us. There are many other transformations in materials that are closely related to this phenomenon. Among them is the spontaneous magnetization of a metal, such as iron for example, when it is cooled from a high temperature under careful conditions such as in the manufacture of magnets. Another example is the transformation of an insulating material into a conductor of electric charges, or even into what is called a superconductor when it does not offer any resistance to the transmission of an electric current. There are many other interesting examples of phase transitions that constitute the most dramatic phenomena in condensed matter physics. Furthermore, the same type of properties have been observed outside of physics, just to give an example, in the behavior of vehicular traffic in cities.

---

*See the references to studies on this topic carried out after 2006 [III.iii.1 to III.iii.4].

Phase transitions consist of a profound alteration of a state of a system and occur drastically when conditions vary, such as, for example, temperature. Often both initial and final states can be observed simultaneously, as in the coexistence of a liquid and its gas separated by a meniscus in a container. The coexistence of these two phases can be controlled by varying the system conditions more carefully, for example by varying both the temperature and the pressure of the fluid so as to maintain the simultaneous presence of the two phases. In this way, the system can be brought to the endpoint of coexistence, which is a special equilibrium state known as "critical point", where the meniscus between the phases disappears and the two states become identical. This critical state has been intensively studied during the last 50 years and a deep understanding of its static and dynamic properties has been achieved. It has very singular properties because in this state fluctuations, or transit between different families of microscopic configurations, occur practically without restrictions.

What is new in our studies is that they focus precisely on only one of these fluctuations, the dominant or most probable fluctuation, made of clusters of molecules, which resemble one of the phases, the liquid or the gas. The critical state is a state of equilibrium that has a fixed average value of its density (in a magnet it would be a zero average value of magnetization). But the fluctuations correspond to a subsystem, a finite region of the whole system, which at a given moment differs from the average density and shows a higher or lower density, like a drop of dense liquid or a bubble of gas. This drop or cluster is unstable, it cannot remain indefinitely, it has a finite life and at a given moment disappears to give rise to a different cluster, and so on. Our results indicate that the time evolution of recurring clusters obeys the same dynamics as the trajectories in nonlinear systems at the transition out of chaos via the intermittency route.

Conclusions

We have briefly described the content of studies carried out by the author and his collaborators on the so-called edge of chaos in nonlinear dynamical systems. Novel properties displayed by simple mappings, already famous due to the knowledge they have contributed to over many years. The studies were motivated by their potential to clarify the limit of validity of an important branch of physics, Boltzmann statistical mechanics. The interest in these studies lies in their ability to illuminate fundamental aspects that may contribute to the resolution of an intense and controversial current debate on the applicability of a generalization of this theory obtained from an iconoclastic modification of the ordinary entropy expression. We also presented candid opinions on these studies and anecdotal aspects of its historical development.

IV.i.1 Génesis de una nueva física estadística. A. Robledo. En Descubrimientos y Aportaciones Científicas y Humanísticas Mexicanas en el Siglo Veinte, Octavio Paredes López, Sergio Estrada Orihuela, editores, Academia Mexicana de Ciencias/ Fondo de Cultura Económica, México, pp. 818-829 (2006). ISBN: 9789681686345

IV.ii How, why and when is q-statistics pertinent

The Renormalization Group (RG) fixed-point maps [IV.ii.1] can be seen as discrete-time versions of the Landau kinetic equation (traditionally used in statistical-mechanical studies of kinetics of phase change [II.iii.1]. In this context these maps are shown to be associated with a Lyapunov function given by the Tsallis entropy. The monotonic iteration time evolution of this function obeys q-statistics and displays q-exponential partition function configuration weights. Also, we have shown [IV.ii.1] that the trajectories of the RG fixed-point maps for the period-doubling and the quasi-periodic routes to chaos can in each case be obtained analytically in a closed form equivalent to the q-exponential expression for the intermittency route. As we explain [IV.ii.1], this requires specific sets of iteration time changes of variable.

Scheme of derivation in IV.ii.1

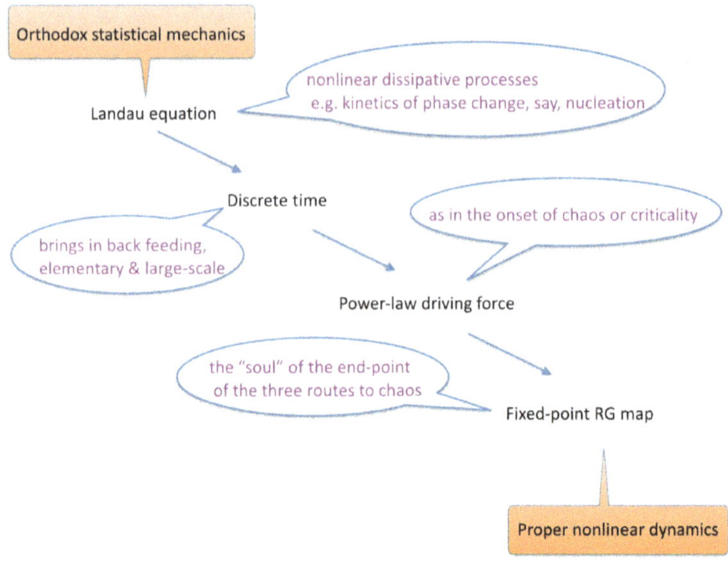

A straightforward mathematical explanation

The probability distributions of ordinary equilibrium statistical mechanics are exponentially decaying functions. Prototypical cases are the canonical ensemble distribution characterized by the Boltzmann Factor and the Maxwell distribution of particle velocities. As a balance to this, the expressions that link the partition functions with the extensive thermodynamic potentials are made through the inverse function, the ordinary logarithmic function. To our knowledge, the mathematical effect in the traditional formalism on these two leading-role functions caused by drastically eliminating (or deleting, removing, disconnecting sets of numbers in) their dense domains of definition has not received much consideration. Physically, this action represents a severe impediment (or obstruction, hindrance, blockade) placed on the statistical-mechanical system to access its otherwise normal set of configurations. An extreme case would be to reduce the set of accessible configurations to a subset of vanishing measure with respect to the original one. An upfront exploration of this mathematical operation is to consider two well-known limit expressions for the ordinary exponential and logarithmic functions: $\exp(x) = \lim_{n \to 0}(1 + nx)^{1/n}$ and $\ln(x) = \lim_{n \to 0}(x^n - 1)/n$. Now, we assume that the demolition (vast numbers elimination) of the real number domains of these functions impede the progress of the above limits up to a given finite value of

$n^* = 1 - q$. We arrive immediately at the definitions of the q-deformed exponential $\exp_q(x) \equiv [1 + (1-q)x]^{1/1-q}$ and logarithmic $\ln_q(x) \equiv [x^{1-q} - 1]/(1-q)$ functions, respectively. These are the functions, inverse to each other, in q-statistics, with index q for the canonical distribution function and index $Q = 2 - q$ for the entropy. See [IV.ii.1]. We have repeatedly encountered these expressions in the quantitative results of our nonlinear dynamical models for complex system phenomena that operate at the transitions to chaos in dissipative low-dimensional maps. See Part III. The attractors at the onset of chaos present in these maps capture trajectories originating from their entire domains of definition and end up evolving only in the subset of positions that form these attractors.

Answers to basic questions

"What is the reason for the occurrence of a limit of validity of traditional statistical mechanics?

This limit can be reached when systems undergo processes that bring about a major obstruction to their previous configurations, leaving access only to small numbers of them, like in glass formation where molecules become caged by their neighbors and can mostly only rattle. In the nonlinear map representation of the system, this happens at the transitions out of chaos. When the behavior remains chaotic, there is insufficient impediment and traditional statistical mechanics remains valid" [85].

"How can a map with only one variable describe macroscopic systems with many degrees of freedom?

For the natural phenomena we study, the variable in the nonlinear maps at transitions to (or out of) chaos already represents a thermodynamic variable of a macroscopic system (like density or energy). The properties obtained correspond to the evolution of a thermodynamic quantity in processes undergone by systems composed of large numbers of degrees of freedom; just as the usual Landau equation does for macroscopic processes observed in ordinary condensed matter systems" [85].

"Are there observable phenomena where the Tsallis entropy provides new understanding?

Within condensed matter physics: the formation of glasses, the transformation of a conductor into an insulator, and critical point fluctuations. Regarding complex systems problems: the phenomenon of self-organization and the development of diversity (biological or social, like languages). Also the comprehension of empirical laws, like those relating to the universality of ranked data or the metabolism of plants and animals" [85].

"What do you think is the most interesting example of a transition into chaos in a

complex system?

My own point of interest is how diverse problems, disconnected at first sight, adopt explanations about different concerns based on a common theoretical formalism or approach. Perhaps a challenging example is that of self-organization, observed in nature but perceived to run against the laws of thermodynamics. The aim was to design a simple model that could be 'run' like a computer program and observe an unpretentious instance of self-organization. The period-doubling cascade into chaos provided one answer" [86].

"Do you think your results could help Tsallis entropy to become a more widely accepted theory?

All proposed scientific advances permeate our community when actual evidence and understanding is provided. This is particularly true for the physical sciences for which eventually available formal proofs and verifiable quantitative predictions are requested. The most interesting episodes are perhaps those where definite answers are not obvious and a saga develops before solutions appear from unexpected places. Places that often were first developed, in say, mathematical themes, with other motivations in mind, or in other cases only recently established" [86].

"How did you first recognize the need to better understand complex systems at a phenomenological level?

I think this has been evident already for some time to many colleagues in the field. There exist large sets of available real data associated with many different complex systems in various disciplines. In contrast, there is absence of comprehensive laws other than a few of empirical origin. The attempts for advancing the understanding of general patterns of behavior have been based on a range of methodologies that imply different mechanisms, while universal principles have remained basically as guiding paradigms, such as the 'edge of chaos' or 'criticality' " [86].

IV.ii.1 How, Why and When Tsallis Statistical Mechanics Provides Precise Descriptions of Natural Phenomena. A. Robledo, C. Velarde. *Entropy* **24**, 1761 (2022). https://doi.org/10.3390/e24121761

Summary and outlook

We have summarized a large catalogue of contributions on stochastic processes, statistical physics, and nonlinear dynamics with applications to condensed-matter physics and to complex system phenomena common to various disciplines. The contributions were created and elaborated basically in Mexico (except for sabbatical periods and other causes) over the last fifty years. The studies have been grouped into thirtytwo sets and these, in turn, into four parts. The order of presentation is not quite chronological as it records reappearances due to unanticipated opportunities to crosslink methodologies that led to interesting analogies. We refer to some of these below, as well as to a few other present ongoing investigations.

Stochastic Processes. Our early works on random walks in the 70's and early 80's [I.i] became most useful in the late 90's, when, via the generalization of the Weierstrass walk to an infinite family of walks, it became apparent that this famous walk is the nontrivial fixed point of the Renormalization Group (RG) transformation devised to operate on this family [I.iv]. More importantly, this fact served as the means to show that entropy optimization had remained an unseen but significant tool for the RG method. Later, [III.iv.5], we showed that continuous time random walks could be combined with the nonlinear dynamics of deterministic intermittency, to provide an unforeseen view of anomalous diffusion. Another random walk process advance, still to be fully developed, is to recast our results for sums of positions at chaotic-band attractors [II, II.iv.1-II.iv.5] in terms of correlated walks. Lastly, we mention our work on renewal processes [I.v], where the expressions for the event probabilities, those for their generating functions, and those for their Laplace transforms, comply with the properties of customary partition functions and ensemble equivalence of equilibrium statistical mechanics. This establishes a strict analogy that links each thermal system model with each renewal process. Since the ideal gas maps into the case of independent events, each particle-interacting model describes a renewal process with correlated events [I.vi].

Density Functional Theory. The order parameter spatial shape of the dominant fluctuation at a critical point can be explored via density functional theory just as it is the case for equilibrium inhomogeneities, say, interfacial profiles or confined fluids. The square gradient approximation is suitable for this purpose [III.i]. The difference with the standard applications is that the fluctuation is unstable and has necessarily a finite lifetime. Its stability can be studied via the 2nd variation of the density functional and its time evolution via the corresponding Landau differential equation [III.i.4]. This procedure makes contact with a nonlinear dynamical problem, that of intermittency close to a tangent bifurcation [3]. Another interesting crosslink with the renewal processes studied in Part I is to consider correlated renewal events via Widom's insertion method [18]. The translation of a showcase thermal system, the so-called Hamiltonian Mean Field Model leads to a renewal process that undergoes a phase transition along time [I.vi]. Yet another instance involves the bifurcation pattern we found in our treatment of hard-core models under the Bethe approximation [II.ii.3], consisting of the instability of the uniform-fluid solution at the highest density, the close-packing density. There, nontrivial oscillatory solid-like solutions appear for the 1st time. A robust analogous behavior can be found in the nonlinear dynamical model for epidemic spreading [78]. This opens the possibility of treating such complex system problems, like obesity [79], with Widom's particle-addition statistical-mechanical technique.

Nonlinear Dynamics. A number of ongoing studies associated with our research lines on nonlinear dynamics have been just described and explained [54]. In addition to them we mention three sets of the most recent investigations. First, we explore new access to statistical-mechanical viewpoints from the known dynamical behavior of the quadratic map attractors by considering the alternative description via probability densities of ensembles of trajectories instead of the trajectories themselves [III.x.1]. Through this we observe the equivalence between the two main paradigms for the understanding of complex systems in recent decades: edge of chaos and criticality. Second, for some time now we have represented the (master) trajectory at the period-doubling onset of chaos in logarithmic scales [III.iv.3]. This representation reveals an infinite family of straight lines, or interlaced power laws, that have led to many fruitful interpretations and analytical statistical-mechanical results [52, 53, 54]. This feature has become now a pathway to demonstrate a common link between all renormalization-group fixed-point maps at the transitions to chaos, intermittency, period doubling and quasiperiodicity. Third, the control parameter gaps that interrupt the chaotic attractor intervals in the bifurcation diagram of the quadratic map contain infinite families of reproductions of the bifurcation diagram itself, the gaps display power-law spacing and widths [71]. These features are caricature building blocks for current modeling of nested complex systems.

The above brief commentaries serve as indications for some feasible short term extensions of the long footpath described here. But also, given the potentially significative general results described in Part IV, there is a wider road to enter and engage in: a limit of validity of ordinary statistical mechanics and the consequent prevalence of a generalized statistical mechanics when conditions apply and place phenomena beyond this limit.

Acknowledgments

AR is profoundly thankful to all his collaborators over the years and deeply appreciative to all colleagues with whom valuable discussions took place. Support is acknowledged from IN106120-PAPIIT-DGAPA-UNAM and 39572-Ciencia-de-Frontera-CONACyT.

Bibliography

[1] H.E. Stanley, *Introduction to Phase Transitions and Critical Phenomena,* Oxford University Press, 1971. ISBN10 0195053168

[2] M.E. Fisher, Renormalization group theory: Its basis and formulation in statistical physics, *Rev. Mod. Phys.* **70**, 653-681 (1998). https://doi.org/10.1103/RevModPhys.70.653

[3] H.G. Schuster, *Deterministic Chaos. An Introduction,* 2nd Ed. (VCH Publishers, Weinheim, Germany, 1988). ISBN-10: 3527293159.

[4] E.W. Montroll and M.F. Shlesinger (1983). Maximum Entropy Formalism, Fractals, Scaling Phenomena, and 1/f Noise: A tale of Tails. *J. Stat. Phys.* **32**:209-230. https://link.springer.com/article/10.1007/BF01012708

[5] P.G. de Gennes, Simple Views on Condensed Matter, Series in Modern Condensed Matter *Physics* **8**, World Scientific 1998 ISBN: 9789810232702

[6] Statphys Conferences, https://en.wikipedia.org/wiki/Statphys

[7] STATPHYS 21 Proceedings, *Physica A*, **306**, pp 1-424 (1 April 2002), A. Robledo, M.C. Barbosa, eds.; https://doi.org/10.1016/S0378-4371(02)00652-0

[8] E.W. Montroll, G.H. Weiss. Random Walks on Lattices. II *J. Math. Phys.* **6** 167 (1965); https://doi.org/10.1063/1.1704269

[9] E.W. Montroll Random Walks on Lattices. III. Calculation of First-Passage Times with Application to Exciton Trapping on Photosynthetic Units *J. Math. Phys.* **10** 753 (1969); https://doi.org/10.1063/1.1664902

[10] B.D. Hughes, M.F. Schlesinger, and E.W. Montroll (1981). Random Walks with Self Similar Clusters. *Proc. Natl. Acad. Sci.* **78**, 3287–3291. https://europepmc.org/article/med/16593023

[11] J.K. Percus, G.J. Yevick. Analysis of Classical Statistical Mechanics by Means of Collective Coordinates. *Phys. Rev.* **110** 1 (1958). https://doi.org/10.1103/PhysRev.110.1

[12] L. Dobrzynski, D.L. Mills. Effect of Reconstruction on the Electronic Free Energy of a Simple Model of Transition Metals. *Phys. Rev. B* **7**, 2367 (1973). https://doi.org/10.1103/PhysRevB.7.2367

[13] E.W. Montroll, G.H. Weiss (1965). Random Walks on Lattices. II *J. Math. Phys.* **6**:167. https://doi.org/10.1063/1.1704269

[14] K. Lindenberg, V. Seshadri, K.E. Schuler. G.H. Weiss. Lattice random walks for sets of random walkers. First passage times, *J. Stat. Phys.* **23**, 11 (1980). https://link.springer.com/article/10.1007/BF01014427

[15] M. Antoni, S. Ruffo, Clustering and relaxation in Hamiltonian long-range dynamics. *Phys. Rev. E* **52**, 2361 (1995). https://doi.org/10.1103/PhysRevE.52.2361

[16] V. Latora, A. Rapisarda, S. Ruffo, Chaos and statistical mechanics in the Hamiltonian mean field model. *Physica D* **131**, 38 (1999). https://www.sciencedirect.com/science/article/abs/pii/S0167278998002176

[17] B. Widom, Some Topics in the Theory of Fluids. *J. Chem. Phys.* **39** (1963) 2808; https://doi.org/10.1063/1.1734110

[18] B. Widom, Structure of interfaces from uniformity of the chemical potential. *J. Stat. Phys.* **19** (1978) pp. 563–574. https://doi.org/10.1007/BF01011768

[19] D.J. Bukman, A.B. Kolomiesky, B. Widom, Fluctuations in the structure of interfaces. *Colloids and Surfaces A* **128** Issues 1–3, 1 August 1997, pp. 119–128; https://doi.org/10.1016/S0927-7757(96)03913-1

[20] J.M. Kosterlitz and D.J. Thouless, Ordering, metastability and phase transitions in two-dimensional systems. Part of Topological quantum numbers in nonrelativistic physics. Published in: *J. Phys. C* **6** (1973) 1181–1203. https://doi.org/10.1088/0022-3719/6/7/010

[21] R. Evans, The nature of the liquid-vapour interface and other topics in the statistical mechanics of non-uniform, classical fluids. *Advances in Physics* **2** 2, 143-200 (1979). https://doi.org/10.1080/00018737900101365

[22] J.S. Rowlinson and B. Widom, Molecular Theory of Capillarity. (Oxford University Press, Oxford 1989) (reprinted by Dover, New York 2002) ISBN-10 0486425444

[23] Y. Rosenfeld, Free-energy model for the inhomogeneous hard-sphere fluid mixture and density-functional theory of freezing. *Phys. Rev. Lett.* **63**, 980 (1989). DOI:https://doi.org/10.1103/PhysRevLett.63.980

[24] P. Tarazona, J. Cuesta, Y. Martínez-Ratón (2008) Density Functional Theories of Hard Particle Systems. In: Mulero Á. (eds) Theory and Simulation of Hard-Sphere Fluids and Related Systems. Lecture Notes in Physics, vol 753. Springer, Berlin, Heidelberg. https://doi.org/10.1007/978-3-540-78767-9_7

[25] V.L. Ginzburg and L.D. Landau, On the Theory of superconductivity. *Zh. Eksp. Teor. Fiz.* **20**, 1064 (1950). https://doi.org/10.1016/B978-0-08-010586-4.50078-X

[26] J W. Cahn and J.E. Hilliard, Free Energy of a Nonuniform System. I. Interfacial Free Energy. *J. Chem. Phys.* **28**, 258 (1958). https://doi.org/10.1063/1.1744102

[27] J W. Cahn and J.E. Hilliard, Free Energy of a Nonuniform System. III. Nucleation in a Two-Component Incompressible Fluid. *J. Chem. Phys.* **31**, 688 (1959). https://doi.org/10.1063/1.1730447

[28] H. Metiu, K. Kitahara, and J. Ross, Derivation of stochastic equations for nonequilibrium Ising mean field model. *J. Chem. Phys.* **63**, 5116 (1975). https://doi.org/10.1063/1.431319

[29] H. Metiu, K. Kitahara, and J. Ross, Stochastic theory of the kinetics of phase transitions. *J. Chem. Phys.* **64**, 292 (1976). https://doi.org/10.1063/1.431920

[30] H. Metiu, K. Kitahara, and J. Ross, A derivation and comparison of two equations (Landau-Ginzburg and Cahn) for the kinetics of phase transitions. *J. Chem. Phys.* **65**, 303 (1976). https://doi.org/10.1063/1.432779

[31] B. Widom, Interfacial tensions of three fluid phases in equilibrium. *J. Chem. Phys.* **62**, 1332 (1975). https://doi.org/10.1063/1.430642

[32] B. Widom, Antonoff's Rule and the Structure of Interfaces near Tricritical Points. *Phys. Rev. Lett.* **34**, 999 (1975). https://doi.org/10.1103/PhysRevLett.34.999

[33] B. Widom, Structure of the $\alpha\gamma$ interface. *J. Chem. Phys.* **68**, 3878 (1978). https://doi.org/10.1063/1.436237

[34] J.W. Cahn, Critical point wetting, *J. Chem. Phys.* **66**, 3667 (1977). https://doi.org/10.1063/1.434402

[35] C. Ebner and W.F. Saam, New Phase-Transition Phenomena in Thin Argon Films, *Phys. Rev. Lett.* **38**, 1486 (1977). https://doi.org/10.1103/PhysRevLett.38.1486

[36] H. Nakanishi and M.E. Fisher, Multicriticality of Wetting, Prewetting, and Surface Transitions. *Phys. Rev. Lett.* **49**, 1565 (Published 22 November 1982). https://doi.org/10.1103/PhysRevLett.49.1565

[37] J.C. Wheeler and B. Widom, Phase transitions and critical points in a model three-component system. *J. Am. Chem. Soc.* **90**, 3064 (1968); https://doi.org/10.1021/ja01014a013

[38] J.C. Wheeler and B. Widom, Lattice-gas model of amphiphiles and of their orientation at interfaces. *J. Phys. Chem.* **88**, 6508 (1984). https://doi.org/10.1021/j150670a012

[39] S.T. Chui, and J.D. Weeks, Phase transition in the two-dimensional Coulomb gas, and the interfacial roughening transition. *Phys. Rev. B* **14**, 4978 (1976). https://doi.org/10.1103/PhysRevB.14.4978

[40] D. Furman, S. Dattagupta, R.B. Griffiths. Global phase diagram for a three-component model. *Phys. Rev. B*, **15**, 441 (1977). https://doi.org/10.1103/PhysRevB.15.441

[41] M. Corti, W. Degiorgio, Critical Exponents near the Lower Consolute Point of Nonionic Micellar Solutions. *Phys. Rev. Lett.* **55**, 19 (1985) 2005. https://doi.org/10.1103/PhysRevLett.55.2005

[42] M.E. Fisher, Long-Range Crossover and "Nonuniversal" Exponents in Micellar Solutions. *Phys. Rev. Lett.* **57**, 15 (1986) 1911. https://doi.org/10.1103/PhysRevLett.57.1911

[43] R.E. Goldstein, "Aspects of the Statistical Thermodynamics of Amphiphile Solutions", Les Houches course Physics of Amphiphiles, Eds. J. Meunier, D. Langevin y N. Boccara, *Springer Proceedings in Physics* **21**, p.261, Berlin 1987. ISBN: 978-1-4613-8389-5

[44] R. Evans, U. Marini Bettolo Marconi, and P. Tarazona, Fluids in narrow pores: Adsorption, capillary condensation, and critical points. *J. Chem. Phys.* **84**, 2376 (1986). https://doi.org/10.1063/1.450352

[45] G.M. De Luca, G. Ghiringhelli, C.A. Perroni, V. Cataudella, F. Chiarella, C. Cantoni, A.R. Lupini, N.B. Brookes, M. Huijben, G. Koster, G. Rijnders & M. Salluzzo, Ubiquitous long-range antiferromagnetic coupling across the interface be-

tween superconducting and ferromagnetic oxides. *Nat Commun* **5**, 5626 (2014). https://doi.org/10.1038/ncomms6626

[46] E.J. König, A. Levchenko, I.V. Protopopov, I.V. Gornyi, I.S. Burmistrov, and A.D. Mirlin, Berezinskii-Kosterlitz-Thouless transition in homogeneously disordered superconducting films. *Phys. Rev. B* **92**, 214503 (2015). https://doi.org/10.1103/PhysRevB.92.214503

[47] M.A.F. Sanjuán, Lord Robert May of Oxford, *Revista Española de Física* **34-2**, 47 (2020). https://revistadefisica.es/index.php/ref/article/view/2662

[48] R.M. May, Simple Mathematical Models with Very Complicated Dynamics. *Nature* **261**, 459-467 (1976). https://www.nature.com/articles/261459a0

[49] T.-Y. Li, J.A. Yorke, "Period Three Implies Chaos", *The American Mathematical Monthly* **82**, 985-992 (1975). https://www.jstor.org/stable/2318254

[50] M.J. Feigenbaum, Quantitative universality for a class of nonlinear transformations, *J. Stat. Phys.* **19**, 25 (1978). https://link.springer.com/article/10.1007/BF01020332

[51] M.J. Feigenbaum, The universal metric properties of nonlinear transformations, *J. Stat. Phys.* **21**, 669 (1979). https://link.springer.com/article/10.1007/BF01107909

[52] A. Robledo, Generalized statistical mechanics at the onset of chaos, *Entropy* **15**(12), 5178 (2013). https://doi.org/10.3390/e15125178

[53] C. Velarde and A. Robledo, Manifestations of the onset of chaos in condensed matter and complex systems, *Eur. Phys. J. Spec. Top.* **227**, 645 (2018). https://doi.org/10.1140/epjst/e2018-00128-9 https://link.springer.com/article/10.1140/epjst/e2018-00128-9

[54] A. Robledo, L.J. Camacho-Vidales, A Zodiac of Studies on Complex Systems, *Rev. Mex. Fís. (Supl.)* **1**, 32 (2020). https://doi.org/10.31349/SuplRevMexFis.1.4.32

[55] N.G. Antoniou et al. Fractals at $T = T_c$ due to Instanton-like Configurations. *Phys. Rev. Lett.* **81**, 4289 (1998). https://journals.aps.org/prl/abstract/10.1103/PhysRevLett.81.4289

[56] Y.F. Contoyiannis and F.K. Diakonos. Criticality and intermittency in the order parameter space. *Phys. Lett. A* **268**, 286 (2000). https://doi.org/10.1016/S0375-9601(00)00180-8

[57] S.A. Marvel, R.E. Mirollo, and S.H. Strogatz, Identical phase oscillators with global sinusoidal coupling evolve by Möbius group action, *Chaos: An Interdisciplinary Journal of Nonlinear Science* **19**, 4 (2009) 043104. https://doi.org/10.1063/1.3247089

[58] C. Tsallis, *Introduction to Nonextensive Statistical Mechanics,* 1st Ed. (Springer-Verlag New York, 2009). ISBN 978-0-387-85358-1. https://doi.org/10.1007/978-0-387-85359-8

[59] U. Tirnakli, C. Beck, C. Tsallis, Central limit behavior of deterministic dynamical systems. *Phys. Rev. E* **75**, 040106(R) (2007). https://journals.aps.org/pre/abstract/10.1103/PhysRevE.75.040106

[60] U. Tirnakli, C. Tsallis, C. Beck Closer look at time averages of the logistic map at the edge of chaos. *Phys. Rev. E* **79**, 056209 (2009). https://journals.aps.org/pre/abstract/10.1103/PhysRevE.79.056209

[61] O. Afsar, U. Tirnakli, Generalized Huberman-Rudnick scaling law and robustness of q-Gaussian probability distributions. *EPL*, 101 (2013) 20003. https://ui.adsabs.harvard.edu/abs/2014PhyD..272...18A/abstract

[62] O. Afsar, U. Tirnakli, Relationships and scaling laws among correlation, fractality, Lyapunov divergence and q-Gaussian distributions. *Physica D*, 272 (2014) 18. https://ui.adsabs.harvard.edu/abs/2014PhyD..272...18A/abstract

[63] H. Jensen, *Self-organized Criticality,* (Cambridge University Press, Cambridge, UK, 1998). Online ISBN: 9780511622717 https://assets.cambridge.org/97805218/53354/frontmatter/9780521853354_frontmatter.pdf https://www.cambridge.org/core/books/selforganized-criticality/0EE3D52D140A543C2F3A706D18CD4F1A

[64] K. Christensen, S. di Collobiano, M. Hall, and H. Jensen, Tangled nature model: A model of evolutionary ecology, *Journal of Theoretical Biology* **216**, 2289 (2002). https://doi.org/10.1006/jtbi.2002.2530

[65] L. Pietronero et al., The uneven distribution of numbers in nature. *Physica A* **293**, 297 (2001). https://doi.org/10.1016/S0378-4371(00)00633-6

[66] M. Kleiber (1932) Body size and metabolism. *Hilgardia* **6**:315–353. https://doi.org/10.3733/hilg.v06n11p315

[67] Kleiber's Law, *Wikipedia*. https://en.wikipedia.org/wiki/Kleiber's_law

[68] G.B. West, J.H. Brown, B.J. Enquist (1997) A general model for the origin of allometric scaling laws in biology. *Science* **276**(5309):122-126. https://doi.org/10.1126/science.276.5309.122

[69] G.B. West, J.H. Brown, B.J. Enquist, The fourth dimension of life; fractal geometry and allometric scaling of organisms. *Science* **284**, 1677-1679 (1999). https://doi.org/10.1126/science.284.5420.1677

[70] J.R. Banavar, T.J. Cooke, A. Rinaldo, A. Maritan. Form, function, and evolution of living organisms, *PNAS* **111** (9) 3332-3337 https://doi.org/10.1073/pnas.1401336111

[71] J.A. Yorke, C. Grebogi, E. Ott, and L. Tedeschini-Lalli, Scaling behavior of windows in dissipative dynamical systems, *Phys. Rev. Lett.* **54**, 1095 (1985). https://doi.org/10.1103/PhysRevLett.54.1095

[72] T. Post, H.W. Capel and J.P. van der Weele. Window scaling in one-dimensional maps. *Physics Letters A* **136**(3) (1989) 109-113. https://doi.org/10.1016/0375-9601(89)90188-6

[73] C. Ross, M. Odell, and S. Cremer, The shadow-curves of the orbit diagram permeate the bifurcation diagram, too. Int. J. Bifurcat. Chaos 19 (2009) 3017. https://doi.org/10.1142/S0218127409024621

[74] C. Tsallis, Possible generalization of Boltzmann-Gibbs statistics. *J. Stat. Phys.* 1988, **52**, 479-487. https://link.springer.com/article/10.1007/BF01016429

[75] A. Cho, A Fresh Take on Disorder, Or Disorderly Science? *Science* **2002**, *297*, 1268–1269. https://doi.org/10.1126/science.297.5585.1268.

[76] C. Tsallis, Enthusiasm and Skepticism: Two Pillars of Science? A Nonextensive Statistics Case. *Physics* 2022, **4**(2), 609-632. https://doi.org/10.3390/physics4020041

[77] Lyapunov function. [Available Online: (accessed on 9 October 2022), https://doi.org/en.wikipedia.org/wiki/Lyapunov_function.

[78] A.L. Hill, D.G. Rand, M.A. Nowak, and N.A. Christakis, Infectious disease modeling of social contagion in networks, *PLoS Comp. Biol.* **6**(11), (2010). https://doi.org/10.1371/journal.pcbi.1000968

[79] E. Lozano-Ochoa, J.F. Camacho and C. Vargas-De-León, Qualitative Stability Analysis of an Obesity Epidemic Model with Social Contagion, *Discrete Dynamics in Nature and Society* **2017**, e1084769, 12 pp, https://doi.org/10.1155/2017/1084769

[80] A. Robledo, "Critical attractors and q-statistics" in *Nonextensive Statistical Mechanics: New Trends, New Perspectives*, JP Boon y C. Tsallis, editors, *Europhysics News* **36**, 214-218 (2005). https://doi.org/10.1007/978-3-540-78961-1_2

[81] *Non-extensive Entropy –Interdisciplinary Applications*, Oxford University Press, C. Tsallis, M. Gell-Mann, editors (2004). ISBN-13: 978-0195159769

[82] See a complete bibliography at http://tsallis.cat.cbpf.br/biblio.htm

[83] J. Havrda and F. Charvat, Quantification method of classification processes. Concept of structural α-entropy, *Kybernetika* **3**, 30 (1967). https://www.kybernetika.cz/content/1967/1/30/paper.pdf

[84] M. Schroeder, *Fractals, Chaos, Power Laws: Minutes from an Infinite Paradise* (Freeman, New York 1991). ISBN: 978-0-486-47204-1

[85] Robledo, A. (2023). Obtaining Tsallis entropy at the onset of chaos. *Research Outreach*. https://doi.org/10.32907/RO-137-5003845529

[86] Robledo, A. (2023). Exploring transitions to chaos in complex systems. *Research Outreach*. https://doi.org/10.32907/RO-137-5035369210

9 798892 489430